Foreign Direct Investment, Ownership, and the Transfer of Technology

European University Studies
Europäische Hochschulschriften
Publications Universitaires Européennes

Series V
Economics and Management

Reihe V Série V
Volks- und Betriebswirtschaft
Sciences économiques, gestion d'entreprise

Vol./Bd. 3085

PETER LANG
Frankfurt am Main · Berlin · Bern · Bruxelles · New York · Oxford · Wien

Ralph Leonhardt

Foreign Direct Investment, Ownership, and the Transfer of Technology

PETER LANG
Europäischer Verlag der Wissenschaften

Bibliographic Information published by Die Deutsche Bibliothek
Die Deutsche Bibliothek lists this publication in the Deutsche Nationalbibliografie; detailed bibliographic data is available in the internet at <http://dnb.ddb.de>.

Zugl.: München, Univ., Diss., 2003

D 19
ISSN 0531-7339
ISBN 3-631-52558-3
US-ISBN 0-8204-7302-2

Printed in Germany 1 2 3 4 5 7

www.peterlang.de

To my parents.

Preface

The first words to be read are written last. It is the author's perpetual curse that the acknowledgments must be an oxymoron.

I am very indebted to Professor Rady, Ph.D., Professor Dr. Schnitzer, Professor Klasen, Ph.D., J. Wolff, Ph.D., Dr. Th. Müller and A. Garcia y Pascual, Ph.D. Their suggestions, encouragement and corrections helped to shape this thesis.

The major part of this work was written during my participation in the research project " German Foreign Direct Investment in Central and Eastern Europe". It was funded by the Deutsche Forschungsgemeinschaft, thus the financial support is gratefully acknowledged.

In particular, I want to express my gratitude to the project chairman Professor Dr. Marin who gave me the chance to participate in this project as the database architect and group leader of the data entry team and to contribute to it in this way. It was a stimulating and exciting experience to work with all the dedicated and excellent people during this unique project.

During the final stages of the thesis I was generously supported by the management and colleagues of the GFD Gesellschaft für Datenverarbeitung GmbH. Mr. Liesenhoff, Mr. Bartels thank you for your appreciation when the accomplishment of this work required me to leave the office at strangest times.

My parents endorsed me all the time, morally and financially. For their perpetual support I dedicate this thesis to them.

Finally, $10^{10^{10}}$ *bisous* for Eva...

München/Munich, Mai 2004

Ralph Leonhardt

Contents

1 Introduction **12**

2 The econometrics of ownership shares **15**

2.1 Introduction 15

2.2 Theoretical framework–Sovereign risk and spillover 19

2.2.1 The model 19

2.2.2 Determinants of ownership shares–Predictions . . . 21

2.3 Estimation 24

2.3.1 Classical models for share equation estimation . . . 24

2.3.1.1 Linear share models 25

2.3.1.2 Nonlinear share models 27

2.3.1.3 Stochastic specifications 28

2.3.2 Subsample regression 29

2.3.3 Conditional Tobit estimation 30

2.3.4 Bias 35

2.3.5 Interpretation of estimates 35

2.3.6 A pseudo-R^2 for the conditional Tobit model 36

2.4 Data and estimation Results 37

2.4.1 Data 37

2.4.2 Variables 37

2.5 Discussion 39

2.6 Conclusion 44

3 International technology transfer and the joint venture life-cycle **46**

3.1 Introduction 46

3.2 Review of the literature 49

3.2.1 Empirical literature 49

3.2.1.1 Firm policy and joint venture instability (Franko 1971) 49

3.2.1.2 The risk of joint venture breakdown (Kogut 1989) 50

3.2.2 Theoretical literature 53

3.2.2.1 Cooperation and rivalry (d'Aspremont and Jacquemin, 1988) 55

3.2.2.2 The pioneer (Kabiraj 1999) 57

3.2.2.3 Local participation requirements and the effect of governmental policy (Sinha 2001) 61

3.2.2.4 Mutual synergies (Roy Chowdhury and Roy Chowdhury 2001) 64

3.2.3 Summary and conclusion 70

3.3 Optimal ownership in international joint ventures and technology transfer–the static case . . . 73

3.3.1 The economic environment 73

3.3.2 Definitions 75

3.3.3 The static model of international joint ventures . . 77

3.3.3.1 Model solution and comparative statics . . 85

3.4 Optimal ownership in international joint ventures and technology transfer–the dynamic case . 90

3.4.1 Introduction 90

3.4.2 Notation 91

3.4.3 Ownership dynamics with autonomous spillover 92

3.4.4 Ownership dynamics with controllable spillover . . 98

3.4.4.1 Introduction 98

3.4.4.2 The indirect Euler equation approach . . . 98
3.4.4.3 On the optimal technology path 102
3.4.4.4 Joint venture breakdown–The transversality condition 106
3.4.5 Spillover and joint venture instability 115
3.5 Discussion 117
3.5.1 Ownership and spillover 117
3.5.2 Aggregate ownership patterns 118
3.5.3 Technology spillover in foreign direct investment . . 123
3.5.4 The value of transferred knowledge 125
3.5.5 Policy recommendations 127
3.6 Conclusion 129

4 Concluding remarks **131**

A Maximum likelihood theory and numerical optimization **133**
A.1 Maximum likelihood theory 133
A.2 Numerical optimization 134
A.3 Implementation issues 135
A.4 Implementation 136

B Static Model–Extensions **144**
B.1 The validity of the first order approach 144
B.2 Market structure and joint venture formation 145
B.3 Joint venture as a truth-revealing mechanism 149

C Dynamic optimization **151**
C.1 Auxiliary lemmata and results 151

D Selected Symbols and Abbreviations **153**
D.1 Selected symbols and abbreviations for chapter 2 153
D.2 Selected symbols and abbreviations for chapter 3 154

vations of ownership dynamics of international investment projects in a unified modeling framework.

The phenomenon of endogenous formation and breakdown of joint ventures is explained as a result of specifics of the technology flow between the joint venture partners. The indirect Euler equation approach is introduced as the mathematical device to derive and analyze the solution of the model.

Several predictions and applications are derived from the model. The model may explain the mixed evidence that was observed empirically when the role of foreign direct investment as a technology transfer channel was explored in the literature. Another application explores the aggregate inter-temporal pattern of ownership in foreign direct investment projects. The model provides a method to measure the value of knowledge transferred by spillover. This measurement concept is applied to data of foreign direct investment in central and eastern Europe.

Particular attention is devoted to the transition economies in central and eastern Europe (CEE). This results from the availability of a microeconomic dataset collected 1997-2002 considering German firms that invested in the transition economies between 1989 and 2001. For an extensive reference see (Universität München 2002). The shock that these countries hit with the introduction of market economies allows the study of economic adaption processes in relative isolation. The existence of a starting point for a market economy is exploited in the discussion of technology transfer via foreign direct investment.

Final remarks and ideas for further research finish this thesis.

Chapter 2

The econometrics of ownership shares

2.1 Introduction

A foreign direct investment, FDI for short, makes a firm a multinational enterprise. In other words, a multinational firm is nothing more than a geographically diversified firm with organizational units that happen to be separated from the headquarter by national borders. These units abroad are just the result of this investment.

The most basic decision the investor has to make with respect to these investment projects is the question of control. The theory of contractual incompleteness of the firm[1] identifies the question of control with the question of ownership because the owner has always the residual rights of his firm.

As a foreign subsidiary usually operates in an unfamiliar environment one might suspect that the investor has a strong interest to keep in control of the investment project as close as possible. In other words, the investor is expected to keep the foreign subsidiary as a whole.

On the other hand, a non-negligible part of investment projects in foreign countries is shared with partners. For instance, for US-based invest-

[1]See, for example, (Grossman and Hart 1986).

ment projects (Desai, Foley, and Hines Jr. 2001), Table I reports that between 19 and 27 % of all investment projects in the years 1982-1997 were shared with local firms. For Germany-based investment projects in manufacturing in central and eastern Europe in 1990 the Bundesbank reports that 94 % of all investment projects were shared!

Why do firms give up a part of their ownership in subsidiaries located abroad, even if these require a high degree of control? What are the theoretical foundations for the determination of ownership in a foreign direct investment project?

Three theories dominate the agenda on this topic. These are

- the transaction cost theory,
- the bargaining theory,
- and the sovereign risk theory.

These theories are partly competing and partly complementary.

The theory with the broadest and longest tradition is the transaction cost theory. It bases on the idea that the intangible assets of a firm are costly to exchange on the market or by an arm's-length contract because of the potential opportunistic behavior of the trading partner. For a review of this broad literature see (Caves 1982) and (Markusen 1995).

The bargaining theory, as set out in (Vernon 1971) and, more rigorous, in (Svejnar and Smith 1984), explains the distribution of ownership shares as the outcome of a bargaining process between the investor and its local partners. The bargaining power determines the size of the ownership share the investor can keep in his hands. The bargaining power is assumed to rise with certain assets the investor can provide and their importance for the success of the investment project. It decreases accordingly with the contributions of the local partners. A recent article that employs this theory and tests it is (Nakamura and Xie 1998).

The last of the three, the sovereign risk theory, is the theoretical base for this chapter. It has a strong and a weak version. The strong version explains the ownership shares as a response to the threat of expropriation of the investment project by the host country's government. This approach was pioneered in (Eaton and Gersovitz 1984).

The more recent, weak sovereign risk theory, as formulated in (Schnitzer 1999) first, explains the ownership distribution as an outcome of the risk of creeping expropriation by taxes and similar governmental measures of the host country.

In this chapter, the weak sovereign risk theory will be empirically tested. Consequently, this approach is complementary to (Nakamura and Xie 1998) that employed the bargaining theory, and to (Asiedu and Esfahani 2001) that integrated the transaction cost and the bargaining theory.

Besides the theoretical base, this chapter complements the aforementioned articles with respect to the econometric techniques employed.

What these two papers, (Nakamura and Xie 1998) and (Asiedu and Esfahani 2001), set apart from the already existing stream of empirical literature in this field is that both pioneered the use of shares as a continuous response variable. This approach is known as the share equation approach[2]. This sharply contrasts to the traditional qualitative response variable approach in this field.

The qualitative response variable technique groups the firms according to the ownership share into groups that are described by a qualitative or ordinal variable. This variable is the response variable in a regression. Depending on the setting, a Logit or Probit estimation or the respective multinomial procedure is applied for testing and inference.

The advantages of the use of the whole range of observations for estimation are evident. While Logit/Probit coefficients are local estimates of the exogenous influences and can be interpreted as marginal effects only, the interpretation in the context of share equations is the same as for the usual OLS estimation. This makes this approach more convenient for interpretation and prediction. Another advantage arises from the avoidance of information loss due to the reduction of the measurement scale. Further-

[2]See (Ronning 1990)

more, because the grouping procedure is a nonlinear transformation and may be subject to some discretion of the researcher results are difficult to compare between studies.

However, while both papers employed basically a sub-sampling approach, where only shares below 100% are used for estimation, this chapter goes one step further. By means of a latent variable approach, a conditional Tobit[3] model is developed. It allows the inclusion of the whole sample for estimation. Increased precision results from this approach. The estimation and testing procedures are derived from maximum likelihood theory.

Another improvement of this approach compared to the aforementioned predecessors is the adaption and inclusion of an appropriate pseudo-R^2. This allows an evaluation of the explanatory value of the estimated model.

As a result of the estimation a surprisingly strong and sustaining *inverse* relationship between profitability and the investor's ownership share can be found. Moreover, the spillover effect of technological knowledge is considered, too. It is found that the less the spillover aversely affects the market position of the investor, the higher is the ownership share that is given away. These two main effects coincide with the theoretical predictions.

The organization of this chapter is as follows. In the next section the theoretical background is given and the hypotheses to be tested are synthesized. This is followed by the description of the data and the presentation of the estimation and testing methods. The results are discussed in the context of the recent literature. It will be finished with conclusions and directions for further research.

[3]It is a reminiscence to the conditional probit model in (Hausman and Wise 1977). I am grateful to Joachim Wolff for coining this term.

2.2 Theoretical framework–Sovereign risk and spillover

2.2.1 The model

The hypothesises to be tested here are derived from the sovereign risk theory of FDI presented in (Müller and Schnitzer 2001). This theory claims, that the investor has an incentive to give up a share of the investment project voluntarily to protect it against the creeping expropriation via taxes by the foreign government.

The article (Müller and Schnitzer 2001) is of particular interest because it established a link to the modern macroeconomic growth theory by the introduction of technology spillover. External effects of technology are a core feature that the theories of endogenous grow share[4].

To get some intuition of the theoretical predictions with respect to the ownership share the model will be outlined here. The predictions are formulated as hypothesises.

Suppose there is an advantageous investment opportunity in a foreign country. The necessary investment to exploit this opportunity requires an initial outlay I. It is assumed that this payment can not be paid by indigenous firms. One might attribute such a situation to a failure in the banking sector and similar restrictions in the capital market as it was often the case in transition economies. This investment I is fixed and sunk.

Given the investment, the investor chooses to give up some share s of his investment project to a local partner, possibly a local firm or government. The remaining part, $1 - s$, is kept by the investor.

After the share distribution is chosen, the government of the host country levies a tax T on the investor's subsidiary.

Now the investor chooses the amount of an additional investment q. This additional investment q is the success probability of the investment

[4]See for example (Romer 1986), (Lucas Jr. 1988), or (Grossman and Helpman 1991), chapter 3 and chapter 4

project. This can be understood as a technology transfer or an additional training of the local workers to improve the success of the investment project. The cost corresponding to a transfer of q is $K(q)$. It is a strictly increasing convex function.

The project return R is realized as a last step. The return is stochastic over $\{0, R\}$ with a probability of success q.

The timing of the model is collected in figure 2.1.

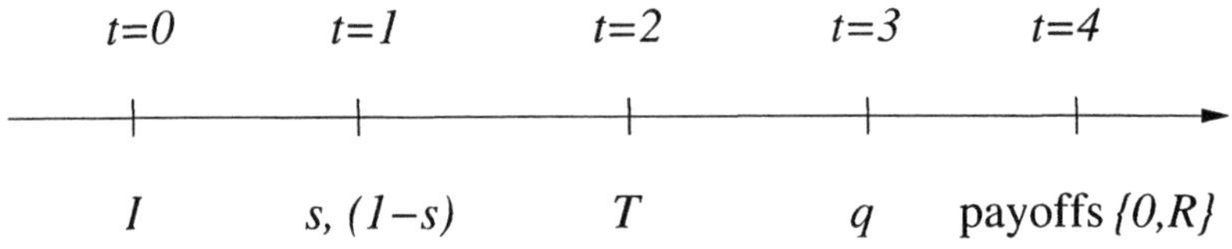

Figure 2.1: Timing of the model

As q is at the discretion of the investor and can be chosen after the government's tax decision, the technology transfer q plays the role of an outside option for the investor in this setup. In case the local government would be too greedy in its taxation decision, the investor reduces the technology transfer q. This in turn reduces the success probability and, if a joint venture was formed, of the spillover.

If the investment is proceeded as a joint venture then a technology spillover S occurs. This spillover benefits directly the joint venture partner but reduces the investor's profits.

The extent of spillover depends on two influences. First, it depends on the transfer decision of the investor. Thus, the success probability q is crucial. If the project fails no spillover can be appropriated.

Second, the ownership share s matters. A higher ownership share increases the spillover. This is reflects the increased accessibility for the joint venture partner to the superior knowledge of the investor.

Nevertheless, the ability of the joint venture partner to capture the spillover may vary. This provides a the reason to introduce an efficiency parameter β. Efficiency is considered from the investor's point of view.

If $\beta < 1$ the spillover generated by the investor is not fully accessible by the joint venture partner. Thus, the partner can capture the investor's loss from spillover only partially. This situation is called efficient. In case $\beta = 1$ the situation is symmetric.

As a last possibility, a situation may arise where the local partner gains more from the foreign knowledge than the loss of the investor, $\beta > 1$. The spillover is denoted inefficient in this case.

From this setup the maximization problem of the investor and the local partner is, respectively:

$$\max_{q,s} U_{Investor} = qs(R-T) - q(1-s)\beta S - sK(q) - I \tag{2.1}$$
$$\max_{T} U_{Partner} = q\left[(1-s)(R-T+S)+T\right] - (1-s)K(q). \tag{2.2}$$

2.2.2 Determinants of ownership shares–Predictions

Now we are ready to discuss the influences of the different decision variables on the ownership distribution.

First, it is restated that the model above has sub-game perfect equilibria with positive investment, positive spillovers and ownership transfer. This provides the foundation for the comparative statics to generate testable prediction.

The results are captured in predictive hypothesises. The reasons are sketched intuitively for every hypothesis. We start with the most important prediction of the model. This hypothesis is unique to the sovereign risk approach. Consequently, it allows a decisive test of the sovereign risk theory.

Hypothesis 2.1 *The more profitable the project the smaller is the multinational's ownership share.*

This is a surprising prediction. Why should the investor give up an increasing share of the investment project if the profit increases?

As the profit rises the incentive rises for the host country's government to capture a rent from the investor by taxes. The spillover becomes less important for the host country to appropriate rents.

The second important prediction deals with the effect of spillover efficiency. To this end, we have to reconsider the basic intuition of the model. Two countervailing forces are balanced by the model. On one hand, there is a government that wants to extract rents from the investment. On the other hand, the technology spillover is considered as important enough to balance the greed of the government against the potential losses if the investor does not transfer enough technology to the FDI project.

The prediction of hypothesis (2.1) plays a particular important role because this prediction is unique to the sovereign risk theory. In contrary, the other theories of ownership in FDI do not provide predictions for the relationship between profitability and ownership.

To characterize the interplay of the ownership decision and technology transfer, we have to consider three situations that are determined by different efficiency regimes of the spillover.

For the case of efficient spillovers, that is $\beta < 1$, there is a rise of the profit of the home country with a decreasing share of the investor. These increased gains are the result of an increased incentive for the investor to increase q under the inefficient spillover. As the spillover is efficient a larger technology transfer does not affect the investor's position as badly as it would do with an inefficient spillover. This yields both a higher expected return of the joint venture as well as a resulting higher spillover S (see equation (2.2)). These benefits outweigh the losses from a reduction of the tax rate T for the host country.

If the spillover is inefficient, $\beta > 1$ this line of arguments ceases. Thus, the incentives of the investor to share ownership vanish. Nevertheless, if the ownership share s is large enough the pressure from the taxation threat may be strong enough to give away some share. Otherwise, it may happen that even in the case of an inefficient spillover, that is the home has additional benefits from the spillover, the home country might chose not to capture a higher share of the investment project. This results from the averse effect that the spillover has on the incentive of the investor to transfer technology. This negative effect turn would reduce the success probability of the project, yielding a reduced expected project profit. The host country prefers to capture the profits of the investor by means of taxes instead.

In the symmetric case, $\beta = 1$, the host country unambiguously benefits from an increased share in the investment project. Although the result for the incentives of the investor is not clear, it can be advantageous again for the investor to give up some share of the investment project to the host country, given that the remaining part of the investment share is large enough for the investor.

The discussion above can condensed into an additional empirical prediction:

Hypothesis 2.2 *The more efficient the spillover the smaller the multinational's ownership share.*

Note that the empirical observations can be used here as a decisive tool to discriminate between different efficiency regimes of the spillover. If the predicted effect in hypothesis 2.3 is observed, even for the less efficient investments, then the investor had enough incentives to share the investment project irrespective of the negative impact the spillover efficiency might have had.

We finish the discussion of the interplay of spillover efficiency and ownership share with less intricate prediction:

Hypothesis 2.3 *The larger the potential spillover S the larger is the multinational's ownership share.*

The notion of potential spillover covers all possible advantages the investor has compared with the abilities of the host country. Of course, to transfer knowledge there must be something that can be transfered. The potential spillover S negatively affects the investor's payoff, while it increases the host country's payoff. Thus, the investor tries to keep the more of the project the more he has to lose by spillover.

The actual burden of the FDI depends on more factors than simply the taxes. These factors that might affect negatively the profitability of the investment project for the investor due to the general political environment are captured by the introduction of political risk. It is emphasized, that this risk is a country specific variable contrasting to a potential deal specific risk. The latter is not considered here.

In reality, there are more means than taxes for the host country to extract rents from an investor. These can be fees and bribes for local authorities, or additional costs due to administrative measures targeted at foreign investors. Even payments to some local *mafia* increase costs considerably [5]

Such influences affect negatively the ability of the investor to keep the ownership of the investment project. Hence, we can state:

Hypothesis 2.4 *The larger the political risk the smaller is the multinational's ownership share.*

2.3 Estimation

2.3.1 Classical models for share equation estimation

Using the ownership share as the response variable leads to an estimation problem that has attracted the interest of researchers for a long time because in many applications shares arise naturally as the variable of interest. Moreover, very often discrete response variables are transformed to shares, because not the particular outcome may be of interest but the average share of a particular outcome compared to the share of all other outcomes. Note that, although share equations seem to be quite restrictive, they are not. Many models can be transformed to a problem that can be formulated in terms of shares. Whenever the response variable is restricted to

[5]From the survey we have anecdotical evidence that there are cases where the local mafia has overtaken some investment projects and sends small amounts of money back to the investors to keep them silent. These were cases from Russia and Romania.

a bounded subset of the real line the structure may be transformed to a share equation[6]. This section gives an overview over standard models for estimation and inference in share equations.

The often used qualitative or ordinal variable models as the result of a grouping procedure of the ownership share is not considered. The implied information loss due to the reduction of the measurement scale is an inadequate procedure if the actual share was observable. An empirical analysis should always employ the maximum of available information. Availability of statistical software for a particular problem class or familiarity of the researcher with particular procedures can not be an argument in favor of procedures that throw away available information.

2.3.1.1 Linear share models

Let s be a vector of observations for the response variable and X be the matrix of independent variable where each row i, that contains the observations x_i for the independent variables, corresponds to a s_i of the response variable. It is assumed that only two categories are observed, that is for every s_i $1 - s_i$ was observed, too. The response variable s is restricted for every observation i,

$$0 \leq s_i \leq 1. \tag{2.3}$$

With a linear share model one gets the usual specification

$$s = X\beta + \epsilon \tag{2.4}$$

with β as the coefficient vector and ϵ as a random vector with expectation zero and a finite variance.

The most basic approach is to assume normality of the disturbances and estimate the coefficients by ordinary least squares. (Maddala 1983) treats this approach under the label 'linear probability model'.

[6]To give an unusual example from robotics, for an autonomous robot with an optical sensor of restricted range that has to follow a moving object the possible ranges are restricted between zero and 180 degree. Thus the response variable is a bounded subset of the real line. The task is to estimate an optimal angle so that the object to follow remains visible.

But this approach ignores several restrictions that are implied by the range constraint of the response variable. The least complicated problem is that the variance of the disturbances can not assumed to be homoscedastic. The heteroscedasticity can be treated in a standard way with a covariance-robust estimator as in (White 1980).

However, if one observes that $s_i = 1$ (or equivalently 100 %) or $s_i = 0$, respectively, the variance must collapse to zero to fulfill the range constraint. The problem of observations that attain values on the limit of the range constraint is not as easily solved. (Greene 1993) proposes in this case to add artificial disturbances to the observations. This idea can be varied in different ways. For example, a shift of the whole observation vector s for a constant value that approaches zero if the number of observations goes to infinity is proposed in (Maddala 1983). In general, the proposed solutions for this problem are not conclusive.

The range constraint in the response variable implicitly constraints the independent variables. In particular, it is required that the moment matrix $X'X$ needs to be positive semidefinite. Otherwise, the $X'X$ would allow for negative values for the predictions which violates the range constraint of the right-hand side. (Ronning and Ringwald 1979) derive conditions for the dependent variable that for any regressor matrix X the moment matrix is positive semidefinite. Of course, one could employ an estimator with side conditions. Using this approach, simplicity is not an argument to chose the linear specification. Moreover, the properties of such estimators are not known in general.

Another complication arises from the fact that shares for the observation i, the shares s_i and s_{-i} must add up to 1 (100 %) if they were observed simultaneously. This adding-up constraint can be automatically fulfilled for some specifications. To this end we need to consider nonlinear share models.

2.3.1.2 Nonlinear share models

Nonlinear specifications currently provide the standard approach in the analysis of share equations.

As already remarked for the simplicity of the exposition only models with two observed categories are considered. Formally, with every observation s_i there corresponds one observation $1 - s_i = s_{-i}$. The estimation equation for both observations s_i and s_{-i} should include the same variables. In other words, category specific variables are not allowed. Using the same variables for the estimation for both categories is important because, under the adding up constraint, only this guarantees the identification of the estimates. We will find two frequently used alternative formulations[7]. One is the translog system model which leads to

$$s_i = \frac{x'\beta_i}{x'\beta_i + x'\beta_i} + \epsilon_i \tag{2.5}$$

$$s_{-i} = 1 - s_i = \frac{x'\beta_{-i}}{x'\beta_i + x'\beta_{-i}} + \epsilon_{-i}. \tag{2.6}$$

This type of models stems from demand analysis where the adding up constraint is well-known, because the expenditures must equal the budget constraint. The translog demand system on which this model is built was proposed first in (Christensen, Jorgenson, and L.J.Lau 1975)[8]. The multiplicative translog system

$$s_i = \frac{x'\beta_i + \epsilon_i}{x'\beta_i + x'\beta_{-i} + \epsilon_i + \epsilon_{-i}} \tag{2.7}$$

$$s_{-i} = \frac{x'\beta_{-i} + \epsilon_{-i}}{x'\beta_i + x'\beta_{-i} + \epsilon_i + \epsilon_{-i}} \tag{2.8}$$

is the other frequently used specification.

These models differ with respect to the treatment of the disturbance terms. In model (2.5) the shares itself are disturbed while in model (2.7) the underlying influences are subject to the disturbances. Again, the incor-

---

[7]This classification follows (Ronning 1990).

[8]See (Deaton and Muellbauer 1980) for an extensive treatment in the context of demand analysis.

poration of the range constraint of the response variable requires additional assumptions. However, for some stochastic specifications of the error terms these assumptions are meet automatically.

2.3.1.3 Stochastic specifications

The discussion of the adequate functional form left open the question how the error terms ϵ should be modeled. A particular appealing specification is the logistic one, because it automatically fulfills the adding-up constraint as well as the range constraint if the multiplicative translog model (2.7) is used. (Considine and Mount 1984) introduced it as the linear Logit share model

$$s_i = \frac{\exp(x'\beta_i + \epsilon_i)}{\exp(x'\beta_i + \epsilon_i) + \exp(x'\beta_{-i} + \epsilon_{-i})} \tag{2.9}$$

$$s_{-i} = \frac{\exp(x'\beta_{-i} + \epsilon_{-i})}{\exp(x'\beta_i + \epsilon_i) + \exp(x'\beta_{-i} + \epsilon_{-i})}. \tag{2.10}$$

Using the well known logit transformation we get

$$\log \frac{s_i}{s_{-i}} = x'(\beta_i - \beta_{-i}) + \epsilon_i - \epsilon_{-i} \tag{2.11}$$

The estimates are identified up to a scale factor. With normal disturbances ϵ_i and ϵ_{-i} this generates a normal disturbance for $\epsilon_i - \epsilon_{-i}$. It follows that the specification (2.11) can be consistently estimated via OLS as long as $s_i > 0$ for every component of s. The Logit approach is the most often used approach for share equations.

Nevertheless, the requirement that all observations for the response variable should be situated in the interior of the constraint set of the response variable is not fulfilled for the dataset (Universität München 2002) used.

This short overview may have given an insight to the currently available toolkit for share equations. It might be noted that shares were always treated as a generic variable, that is, as a direct outcome of the data generation process. For ownership shares this point of view is insufficient. A structural approach to the econometric modeling of share equations might be more appropriate to capture the peculiarities of the data generation process in ownership decisions.

2.3.2 Subsample regression

As mentioned in the introduction, the ownership share is taken as a continuous response variable. This approach was adopted quite recently. Namely the article (Nakamura and Xie 1998) pioneered this approach. A similar effort was undertaken in (Asiedu and Esfahani 2001).

A common feature, that both approaches shared, is that the observations are divided into two subsamples. One subsample contained all observations with $s_i < 100$ (partially owned subsidiaries, POS), the other subsample contained all wholly owned subsidiaries (WOS).

Both applied a probit regression to the dummy response variable to be a WOS or not. It should be emphasized: The variable that selected the WOS subsample served simultaneously as the variable to be analyzed. Such a procedure usually raises severe identification problems.

For the POS sample, in contrast, diverging routes were chosen. Nakamura and Xie employed for the POS sample the two-step Heckman estimator [9] with the WOS-sample as the selection equation. Although there are circumstances that such an approach is valid, from the specification of the aforementioned article it can not concluded that the simultaneity problem was given the attention it is due in this context[10]. We will not discuss the validity of the approach chosen in (Nakamura and Xie 1998) at length herein beyond mentioning that there could arise a problem. First, because their estimation procedure was not fully specified in their article and thus it would be more a guess if they took care for this problem. Second, the conditional Tobit approach developed herein avoids this kind of problems at all and makes these considerations superfluous.

(Asiedu and Esfahani 2001) used a Tobit-model for the POS sample

[9]See (Maddala 1983), chapter 9.

[10](Maddala 1983), chapter 5 surveyed the related problems.

with truncation at 100 %, following an idea in (Cragg 1971). This avoided, of course, the questions of identification that arose in the approach by Nakamura and Xie.

Both articles emphasized that for the WOS-sample not necessarily the same influences are crucial as for the POS-sample. Well, this might be the case and would possibly give sufficient information to identify the WOS-estimates. However, it is emphasized that the use of category-specific variables was not a result of their estimation. It was an assumption to proceed their proposed estimation procedure. Thus, the argumentation line for the WOS-sample was circular.

To see this, assume for the moment that some covariates would go into both estimation equations, i.e. for the WOS and the POS at the same time. If the joint variables would be insignificant one could argue that this shows that these variable are not influential. Certainly, these estimates would be biased if the identification is not guaranteed a priori. Thus, this test is not conclusive. The same applies if these common variables were significant.

To break this circle a totally different approach should be used. It is presented in the next subsection. Nevertheless, a truncated regression similar to (Asiedu and Esfahani 2001) is included for the POS-subsample for comparison purposes. The identification problems are not present in this case. But of course, the potential for incidental truncation is neglected.

2.3.3 Conditional Tobit estimation

Instead of taking the ownership share as a generic variable, it will be considered as the observable outcome of a latent variable. From this starting point the conditional Tobit approach is introduced. Although the latent variable strategy of modeling partially unobservable decision variables of individuals is not new, in the context of share equations it is wholly new idea. And of course this approach can easily be generalized to proportions data where a lot of 0 % or 100 % values are observed.

The paradigm shift realized by this approach is not to treat observations that actually take the values at the limits of the interval in a special way. They are integrated in the estimation and testing process the same way like all other observations are.

In addition, the latent variable approach is quite appealing for the ownership question. While in traditional share equations, for example for market shares, the share distribution itself is the result of the data generation process, the ownership shares are fully at discretion of the firm owner. Consequently, one should explore the determinants of the decision process that resulted in the share distribution instead of a direct exploration of the determinants of the share distribution. A latent variable model provides such an approach.

The investment project is an attempt of the firm to increase its utility. In contrast to consumer theory utility is understood here as a shorthand for expectations of the prospective future capital flow. These, of course, can neither be observed by an outsider nor given some specific parametric form. That's why, it is labeled utility. Accepting this way to capture the expectations of the firm's future profits, the "utility" is cardinal with a given scale. That is we can interpret ratios of differences in utility. Without this assumption the estimation procedures described here would be logically impossible.

The utility u_i that an investment project i generates for the investing firm is assumed to be linear in its influences $x_{ij}, j = 1, \ldots, k$. Additional unknown influences are captured by deviations from the mean $x_i'\beta$ of the utility as usual. These random deviations are caught in an error term ϵ_i for the particular investment project i.

Thus, we can assume the usual regression equation for the latent utility variable u_i:

$$u_i = x_i'\beta + \epsilon_i. \tag{2.12}$$

If this utility is below some level c_l we can not observe any investment. It follows that we have a truncation of the distribution of ownership shares s_i of the investor for all firms that have an utility below c_l.

On the other hand, if the utility is above some level c_u the firm does not give away any share of its investment project. Of course, all firms that have more than this level c_u own the investment project fully.

Formally, the relationship between the ownership share and the utility from investment is expressed as

$$s_i = \begin{cases} \text{no investment, thus no observations at all} & u_i < c_l \\ s_i = u_i & c_l \leq u_i < c_u \\ 100 & c_u \leq u_i \end{cases}.$$

Hence, we have to estimate a linear regression model with left truncation and right censoring. It can be considered as a classical Tobit model[11], that is a model with right censoring, conditional on observing the firm as an investor. This yields the name of this approach as *conditional Tobit estimation.*

Using results on conditional distributions[12], construction of a maximum-likelihood estimator is straightforward. This has the additional advantage that the apparatus of maximum likelihood theory immediately delivers the distribution of estimators[13]. Thus we are left with the adaption of the general theory to the peculiarities of the problem at hand.

The likelihood function for this model under the normality assumption for the error terms, $\epsilon \sim N(0, \sigma^2)$, and independence of the observations is

$$\begin{aligned} \mathcal{L} &= \prod_i \left(1 - \Phi\left(\frac{c_l - x_i'\beta}{\sigma}\right)\right)^{-1} \times \\ & \prod_{\{i:c_l \leq u_i < c_u\}} \frac{1}{\sigma}\phi\left(\frac{u_i - x_i'\beta}{\sigma}\right) \prod_{\{i:c_u \leq u_i\}} \Phi\left(\frac{c_u - x_i'\beta}{\sigma}\right). \end{aligned} \tag{2.13}$$

A comparison of this likelihood function with the likelihood function of the standard Tobit model[14] shows that this conditioning incorporates a

[11] See (Amemiya 1985)

[12] See (Mood, Graybill, and Boes 1974), part IV.

[13] See section A in the appendix

[14] See (Judge, Hill, Griffiths, Lütkepohl, and Lee 1988) section 19.3

penalty for all observations that are strictly positive. Thus, the likelihood function is transformed strictly monotonically and the well behavedness of the standard Tobit likelihood function as stated in Olsson (Olsson 1978) is conserved.

For estimation purposes, the log-likelihood function is to be maximized. Several specialized algorithms for the numerical maximization of the log-likelihood function are available. The details are given in the appendix A. Although they are all equivalent with respect to the results, they vary with respect to the necessary ingredients.

However, all of these algorithms need the gradient of the log-likelihood function. It can be computed to be for the coefficients[15]

$$\begin{aligned}\frac{\partial \ln \mathcal{L}}{\partial \beta_j} &= -\sum_i \frac{\phi(\frac{c_l - x_i'\beta}{\sigma})}{1-\Phi(\frac{c_l - x_i'\beta}{\sigma})} \frac{x_{ij}}{\sigma} + \\ &\sum_{\{i:c_l \le u_i < c_u\}} \frac{u_i - x_i'\beta}{\sigma} \frac{x_{ij}}{\sigma} + \sum_{\{i:c_u \le u_i\}} \frac{\phi(\frac{c_u - x_i'\beta}{\sigma})}{1-\Phi(\frac{c_u - x_i'\beta}{\sigma})} \frac{x_{ij}}{\sigma}\end{aligned} \tag{2.14}$$

and for the variance

$$\begin{aligned}\frac{\partial \ln \mathcal{L}}{\partial \sigma^2} &= -\sum_i \frac{1}{2\sigma^2} \frac{\phi(\frac{c_l - x_i'\beta}{\sigma})}{1-\Phi(\frac{c_l - x_i'\beta}{\sigma})} \frac{c_l - x_i'\beta}{\sigma} + \\ &\sum_{\{i:c_l \le u_i < c_u\}} \left[-\frac{1}{\sigma^2}\frac{1}{2\sigma^2} \left(\frac{u_i - x_i'\beta}{\sigma}\right)^2 \right] + \\ &\sum_{\{i:c_u \le u_i\}} \frac{1}{2\sigma^2} \frac{\phi(\frac{c_u - x_i'\beta}{\sigma})}{1-\Phi(\frac{c_u - x_i'\beta}{\sigma})} \frac{c_u - x_i'\beta}{\sigma}.\end{aligned} \tag{2.15}$$

These derivatives indicate that the Hessian or the Fisher-Information may be cumbersome to derive. It follows that the easiest way to maximize this likelihood function it is the BHHH-Algorithm[16] because it relies on the first derivatives only.

---

[15] The terms are given to support computational simplicity. Of course some terms can be more compressed.

[16] Or BH^3 as it is sometimes referred too

Using the equivalence given in theorem (A.2), it replaces the necessary expectation for the scoring algorithm[17] with its sample equivalent.

This means, we take the average of the outer product of the gradient of the likelihood function for every observation. For this purpose we need the single-observation gradient of the likelihood function.

It is for the likelihood function (2.13) and for the coefficients i and the observation j

$$\frac{\partial}{\partial \beta_i}\mathcal{L}_j = -\frac{\phi(\frac{c_l - x_i'\beta}{\sigma})}{1-\Phi\left(\frac{c_l - x_i'\beta}{\sigma}\right)}\frac{x_{ij}}{\sigma} + \begin{cases} \frac{u_i - x_i'\beta}{\sigma}\frac{x_{ij}}{\sigma}, & \text{if } y_j \in (c_l, c_u] \\ \frac{\phi(\frac{c_u - x_i'\beta}{\sigma})}{1-\Phi\left(\frac{c_u - x_i'\beta}{\sigma}\right)}\frac{x_{ij}}{\sigma}, & \text{if } y_j = c_u \end{cases} \tag{2.16}$$

and for the variance

$$\frac{\partial}{\partial \sigma^2}\mathcal{L}_j = \frac{1}{2\sigma^2}\frac{\phi(\frac{c_l - x_i'\beta}{\sigma})}{1-\Phi(\frac{c_l - x_i'\beta}{\sigma})}\frac{c_l - x_i'\beta}{\sigma} + \begin{cases} \left(-\frac{1}{\sigma^2} + \frac{1}{2\sigma^2}\left(\frac{u_i - x_i'\beta}{\sigma}\right)^2\right), & \text{if } y_j \in (c_l, c_u] \\ \frac{1}{2\sigma^2}\frac{\phi(\frac{c_u - x_i'\beta}{\sigma})}{1-\Phi(\frac{c_u - x_i'\beta}{\sigma})}\frac{c_u - x_i'\beta}{\sigma}, & \text{if } y_j = c_u. \end{cases} \tag{2.17}$$

The drawback of the BHHH approach is that one can not use the concentrated likelihood function to estimate the variance separately from the coefficients. But this is only a minimal disadvantage compared to the cumbersome analytical derivation of the Hessian.

Finally, the distribution of estimates can be derived from theorem (A.1). Note, that the BHHH-algorithm estimates all the necessary parameters simultaneously. Thus, one can directly take the estimates from the optimization procedure and plug them in.

---

[17] See the definition (A.6) in the appendix A

2.3.4 Bias

As the likelihood estimator is consistent one might ask what the bias of the subsample regression is compared to the conditional Tobit approach. Let us denote the expectation of the subsample approach as $E(\hat{Y}_t)$ and the expectation of the conditional Tobit approach as $E(\tilde{Y}_i)$.

The conditional Tobit approach assumes that

$$E(\tilde{Y}) = E(Y|Y \in (c_l, c_u)) + E(Y|Y \geq c_u)\text{Prob}(Y = c_u). \tag{2.18}$$

The subsample regression in contrast assumes that

$$E(\hat{Y}) = E(Y|Y < c_u) + E(Y|Y \geq c_u)\text{Prob}(Y = c_u). \tag{2.19}$$

We have accordingly

$$E(\tilde{Y}_i) - E(\hat{Y}_i) = -E(Y_i|Y_i < c_l)\text{Prob}(Y < c_l). \tag{2.20}$$

As long as we assume that there is a positive probability that some firms have a preference to invest but do not so because the utility level is to small, this implies that the subsample approach overestimates the absolute value of the coefficients. In other words, they are too large for positive values and too small for negative values. This is the conclusion from equation (2.20) that the respective predictions under the different assumptions have a negative difference and the response restriction that positive values are possible only.

2.3.5 Interpretation of estimates

The conditional Tobit model has a clear advantage compared to the usually chosen multinomial Logit approach. The parameter estimates are not marginal effects as in the Logit/Probit approach. They can be interpreted as it is usually done in the classical OLS approach instead.

Moreover, the grouping with respect to the ownership shares that is usually necessary to employ the techniques from a discrete choice approach might lead to a loss in information. The estimation technique employed here avoids this information loss. The conditional Tobit approach, in contrast, employs all data without an arbitrary transformation.

Of course, an approach that employs the whole sample without information reducing transformations is preferable. Nevertheless, from an applied point of view, it turns out that the subsample approach leads to a reasonable good fit for the data. Otherwise, the variance estimator differs considerably for the subsample approach. It yields differences in the testing distributions.

2.3.6 A pseudo-R^2 for the conditional Tobit model

Finally, we are left with the question how one can assess the estimation performance of the estimated model in comparison to a situation where no covariates are used at all.

It is required, that this goodness-of-fit measure should be applicable to the subsample approach as well as to the conditional Tobit estimation.

For this purpose the McKelvey-Zavoina pseudo-R^2 is adapted from the realm of discrete choice models. This extends the idea of (Laitila 1993) to the conditional Tobit estimation. His pseudo-R^2 is constructed as

$$R^2_{MZ} = \frac{\sum_i^N \left(\hat{u}_i - \overline{\hat{u}_i}\right)^2}{\sum_i^N \left(\hat{u}_i - \overline{\hat{u}_i}\right)^2 + N\hat{\sigma}^2}. \tag{2.21}$$

Here $\hat{u}_i$ is the predicted utility for observation i, $\overline{\hat{u}_i}$ is their respective average. The $\hat{\sigma}^2$ denotes the estimated variance. Thus, it expresses in some sense the goodness of fit with respect to the underlying latent variables. After the adaption of the predicted values, one can use this pseudo-R^2 directly.

The advantage of this pseudo-R^2, compared to a likelihood-based measure[18], is that it allows a comparison of estimation models with different likelihood functions. This property is needed here. The pseudo-R^2 avoids

[18](Ronning 1991), section 2.5.1 discusses several goodness-of-fit measures. (Zimmermann and Veall 1996) provides a simulation study of the various measures for the standard Tobit model.

an additional difficulty of likelihood-based measures. As the conditioning of the likelihood function puts a penalty on the likelihood functions, likelihood-based measures of fit always prefer unconditional models. But this would be misleading in our setting.

Note that, given appropriate regularity conditions[19], the R^2_{MZ} converges in probability to the sample R^2 if the limits go to ∞ and $-\infty$, respectively. It coincides with the standard R^2 in this situation.

2.4 Data and estimation Results

2.4.1 Data

The data employed for this estimation stems from a survey of German firms that invested in transition economies from 1989 to 2000. Altogether 380 German firms were covered. These firms had 924 investment cases that grossed to 38 Bio. Deutsch Mark.

A firm was considered to be an investor if it had invested at least 1 Mio. Deutsch Mark, or 0.5 Mio. in case of a service firm, and if it holds at least 20 % of the equity of the investment project. If the investment project was a pure sales branch, it was excluded if the investor was not a pure trading firm.

The relevant variables were extracted from the dataset. If one observation had a missing value for one variable the whole observation was discarded. This leaves 394 observations in the sample. From these 394 observations 194 are wholly owned subsidiaries. That is the upper range constraint is attained. These are 49 % of the employed observations.

2.4.2 Variables

The response variable for the estimation procedures is the ownership share of the investor in the foreign country SHARES. As mentioned above, due

[19] See (Laitila 1993)

to the survey design[20], only investments with an ownership share of the investor of at least 20 % and an investment of at least 1 Mio. Deutsch Mark or at least 500,000 Deutsch Mark for firms in the services sector were observed.

For the evaluation of the political risk in the host country the State Capture Index (SCI) was chosen[21] as reported in the Transition Report 2000 (EBRD 2000). It is a composite measure of different governmental characteristics and grows from 0 for the lowest political risk without a bound along the positive axis. Due to the data source for the SCI only numbers for the year 2000 are available.

Note that, although the questionnaire provided questions about the risk situation in the investment country at the moment of investment, these numbers can not be used to test the theory because the political risk must be country specific instead of investment specific.

The project profitability is measured by the ratio of the sales of the FDI net of the labor costs to the invested sum in the investment project ((ISALES-ILCOST)/FDISUM). It is emhasized that the difference of sales to labor costs is not enough to support the hypotheses (2.1). Without the relation to the invested sum it would be prone to pure size effects. For example, the larger a project the higher is the probability that the investors prefers to share the burden of risk with indigenous firms.

To assess the potential spillover R&D personnel as a share of all employees of the investor is chosen (EMPRNDL/EMPL). This construction deviates slightly from th R&D expenditures to sales ratio that is usually taken as the measure for the technological advantage. The share of R&D personnel to the whole workforce of the parent firm used herein is considered advantageous. First, many firms have employed engineers or scientists for their operations. The wages paid to them are usually not considered as R&D expenditures. Especially small, research oriented firms do not even

[20]This setting was necessary to make the figures comparable to the data that the Bundesbank publishes.

[21]See (Hellmann and Jones 1999) for the definition. I am grateful to Monika Schnitzer for proposing this index.

have a R&D budget although the firm as a whole works in a research related area as microbiology or software development. Thus, the proportion of employees in R&D-related areas might provide a more reliable measure of technological ability of the firm than an expenditure based variable might provide. Second, contrasting to the expenditure based measures of R&D intensity that describe a flow, the measure herein is based on a stock of human capital the firm is able to employ.

The evaluation of spillover efficiency is much more complicated. From theoretical grounds, it should be a variable that reflects on one hand the impact of the FDI project on the host country's market without being correlated to success measures as, for example, sales. This requirement is hardly hit by one proxy variable alone. This is the reason to employ different variables to measure spillover efficiency.

Table (2.1) provides summary statistics for the employed variables. An overview of variables, predicted effects and variable descriptions is given in table (2.2).

Variable Name	Dimension	Mean	1.Quartile	Median	3.Quartile
SHARES	%	78.69	51	95	100
SCI	-	15.03	11.00	12.00	21.00
PROFITAB	%	2.77	0.43	1.38	3.94
POTSPILL	%	12.50	0.80	5.00	13.4
IEXP	%	65.34	10	100	100
EFFSPILL	%	56.29	0.0325	80	100
EFFSPLS	DM	17520000	0	1764000	8000000

Table 2.1: Variable Descriptives

2.5 Discussion

The results of the estimation and the tests are given in table (2.3). As expected, the coefficients for the conditional Tobit model are smaller in

Variable Name	Variable Description	Feature captured	Predicted Sign
	endogenous Variable		
SHARES	ownership share of the investor		
	exogenous Variables		
SCI	State Capture Index	political risk	negative
PROFIT	(Turnover-Labor Cost)/Investment	project profitability	negative
POTSPILL	ratio of R&D personnel to total number of employees of the investor	potential spillover	positive
IEXP	production share sold by the investment project	spillover efficiency	negative
EFFSPILL	share of production sold by the investment on its home market	spillover efficiency	negative
EFFSPLS	value of sales of the investment on its home market	spillover efficiency	negative

Table 2.2: Variables and predicted signs for the model

absolute value than the coefficients for the subsample regressions as the theoretical result for the bias, equation (2.20), predicted.

With respect to the theory to be tested, the estimation results mostly support the theoretical predictions. Nevertheless, there are some valid counter-arguments that need to be considered.

For the political risk variable, SCI, a negative sign is observed as it was predicted by the theory. If the state capture index (SCI) rises, which indicates an increased sovereign risk for the investor, the ownership share of the investor in the investment project decreases. Irrespective of the other included variables SCI is significant and has the largest numerical effect of all other variables. We find that an increase of the state capture index for one decreases the participation share of the investor for a share between

Variable Name	Conditional Tobit I	Subsample-Regression I	Conditional Tobit II	Subsample-Regression II	Conditional Tobit III	Subsample-Regression III
const.	100.24** (30.4)	109.2** (19.02)	101.89** (31.45)	110.27** (18.8)	100.84** (36.03)	107.54** (21.03)
SCI	**-0.535**** [a] (−3.555)	**-0.596*** (−2.22)	**-0.542**** (−3.602)	**-0.605*** (−2.240)	**-0.669**** (−4.501)	**-0.729*** (−2.736)
PROFITAB	**-0.166**** (−4.242)	**-0.180*** (−2.38)	**-0.182**** (−4.704)	**-0.194*** (−2.584)	**-0.175**** (−4.747)	**-0.197*** (−2.728)
POTSPILL	−0.136* (−2.487)	−0.138 (−1.21)	−0.133* (−2.350)	−0.134 (−1.155)	−0.081 (−1.362)	−0.065 (−0.552)
IEXP	**-0.124**** (−4.171)	**-0.138*** (−2.59)				
EFFSPILL			**-0.086**** (−3.01)	−0.097 (−1.858)		
EFFSPLS					**-0.0018**** (−5.714)	**-0.0011**** (−3.824)
Observations	394	205	394	205	394	205
R^2_{MZ}	0.126	0.255	0.255	0.231	0.1133	0.2883

Table 2.3: Estimation Results

[a] * denotes significance of the result at the 5 % level for the coefficients, ** denotes significance at the 1 % level; correctly predicted signs for significant values are in bold fonts

0.535 % and 0.729 % depending on the other variables included.

The most surprising theoretical prediction, the *inverse* relation between project profitability and ownership shares, achieves a strong support, too. This is shown clearly by the significant negative coefficient for the PROFITAB variable irrespective of the constellation of the other variables. The effect of the profitability is a reduction of the ownership share between 0.166 % and 0.197 % with a reduction of profitability for 1 %.

These two observations support the sovereign risk as a dominant influence on the distribution of ownership shares. It is suprising that these relations hold irrespective of the branch or the host country of the investment project. This indicates how strong the effect of these two variables is.

On the downside, the estimates of the POTSPILL variable argue against the theory. The negative sign indicates that the larger the R&D-intensity of the investing firm at home the lower is the share of the investor in the investment project. In terms of the model, an increased potential spillover implies a reduced share of the investor. Nevertheless, the estimates are not always significant. As we will see, the wrong sign might be the result of the strong interdependence between proxies for the potential and the effective spillover.

The spillover efficiency measured in different terms has in contrast to the potential spillover variable the right sign and is significant for 5 of the six estimates.

Table (2.4) provides the correlation coefficients for the independent variables. These correlations might provide an explanation of the failure of the POTSPILL variable to support the theory. The correlations indicate a relatively strong dependence between potential spillover and spillover efficiency. Of course, although the theoretical concepts for the potential spillover and the spillover efficiency are quite different, the measurement of the two concepts has to use variables that arose from similar characteristics of the investment project. They are closely linked accordingly as the correlation table (2.4) suggests.

	SCI	PROFITAB	POTSPILL	IEXP
SCI	1			
PROFITAB	0.044	1		
POTSPILL	-0.035	-0.023	1	
IEXP	-0.047	-0.07	0.013	1
	SCI	PROFITAB	POTSPILL	EFFSPILL
SCI	1			
PROFITAB	0.044	1		
POTSPILL	-0.035	-0.023	1	
EFFSPILL	0.02	0.173	0.143	1
	SCI	PROFITAB	POTSPILL	EFFSPLS
SCI	1			
PROFITAB	0.044	1		
POTSPILL	-0.035	-0.023	1	
EFFSPLS	-0.11	0.045	0.21	1

Table 2.4: Correlation among exogenous variables

Another reason outside this line of argumentation arises from the fact that on average the investors emphasized the market position as most important reason for their investment project. The survey provided a question for the main reasons to invest in central and eastern Europe. The question for the most important reason was scored on a scale from 1 to 5 on average for the option[22] 'Market Entry' 1.90, for 'Market Proximity' 1.93 and for the 'Market size' 2.4. This indicates that the possible loss of intangible assets was not given the attention that the market position attracted. So the considerations concerning the market position of their investment project possibly dominated those concerning the risk of technology loss.

The question which variable for spillover efficiency provides the best model requires a reference to the pseudo-R^2. Given the data is a cross-section, a reasonable R^2 is reported for all models. For the subsample regressions the model which included the EFFSPLS variable gives the best model. The POTSPILL variable is insignificant in this case.

[22]These were the highest average scores for this question.

The inclusion of the EFFSPILL variable is preferable by virtue of a R^2 of 25.5 % for the conditional Tobit estimation. However, in this model the POTSPILL variable is significant with the wrong sign. As the conditional Tobit approach guarantees an unbiased estimation, whereas the sub-sample regression might be biased, the conditional Tobit model should be preferred.

Altogether, the best model is given by the conditional Tobit estimation II which uses EFFSPILL as an estimate for spillover efficiency.

2.6 Conclusion

The purpose of this investigation was the test of hypotheses concerning the ownership share derived from the paper (Müller and Schnitzer 2001). For this purpose a subsample regression approach was adapted and the conditional Tobit method was developed.

The predictions from the sovereign risk theory for the ownership distribution are mostly supported. Nevertheless, the variable describing the spillover potential either bears the wrong sign (for the conditional Tobit model) or is insignificant (for the subsample regression). This might be a result of the interplay between the spillover risk and the effectiveness of the spillover where the latter dominates.

The empirical approach via the conditional Tobit model avoids the grouping of observation as it is required by the multinomial Logit model which is the established technique in the literature on the estimation of ownership shares. The conditional Tobit model is thus considered advantageous compared to the multinomial Logit approach.

However, as this approach demonstrates a better model in theory not implies a better fit of the data. This is, of course, an embodiment of the bias versus variability trade-off.

A more detailed study of this trade-off is a direction that should be explored. Because the particular range constraint that ownership shares entail is a data situation that appears often such a study could be useful even outside the context of ownership shares estimation.

With respect to the employed economic theory, it must be admitted that only a part of the theory was tested here. The theoretical framework developed in Schnitzer and Mueller (Müller and Schnitzer 2001) predicts effects of infrastructure investments on the ownership share, it endogenizes to some extent the political risk (the tax choice), and the extent of the spillovers (given the taxes).

The reason to favor the ownership decision over the other questions arose from the theoretical framework as well as from the importance of the topic of ownership. As recent economic research suggests, the distribution of the ownership share might be the most important decision for an enterprise.

The conditional Tobit approach developed herein gives a new tool at the hand of researchers who seek to test the implications of the recent developments of the theory of the firm and the questions of ownership and control. It is tailored to all kind of proportions data where the boundaries of the response interval actually appear among the observations. As an example, the determinants of ownership share were analyzed.

The usefulness of the conditional Tobit approach in the modeling of share equations depends on the extent of truncation. A Monte-Carlo study of the influence of the truncation on the estimation and testing result could provide hints for improved performance of the procedure in finite samples in detail. The same applies to the adaption of the pseudo-R^2. At least for the data at hand, the estimation results indicate that the influence of truncation can not be neglected. In general, the previously used subsampe approach overestimated the effects of independent variables on the response variable.

The exciting and fruitful theoretical linkage that the theoretical framework established between the theory of spillover and the theory of sovereign risk of foreign direct investment made it a grateful field of empirical investigation. And, as it turned out, the most theoretical results were supported when confronted with reality.

Chapter 3

International technology transfer and the joint venture life-cycle

3.1 Introduction

Modern growth theory assigns technology spillover a prominent role among the determinants of growth in industrialized countries. Without spillover perpetual growth can not be achieved. The involuntary transfer of technology between firms and countries creates the base for sustainable growth.

The monograph (Grossman and Helpman 1991) has integrated modern growth theory with international trade. This integration builds the base for empirical explorations of the relations between trade, growth and spillover. The usually rich data on international activities of firms, exports, investment, and cross-border licensing provides a natural base for the empirical study of these relationships.

The article (Coe and Helpman 1995) shows that foreign technology measured as the stock of R&D expenditures has empirically a significant and important influence on growth. In their approach, export is given the role of a knowledge transfer channel. But if exports have such a strong effect this should be the case for multinational firms created by foreign direct investment, too.

The article (Lichtenberg and van Pottelsberghe de la Potterie 1996) pioneers the exploration of foreign direct investment as a vehicle for spillover. But the results are mixed. This partial failure boosted a lot of empirical studies but the results remained unsatisfactory. (Blomstrom and Kokko 1996) reviews the relevant empirical literature.

Recently, (Xu 2000) finds a positive relationship between outward FDI of US-firms and productivity growth in the host country for the years 1966-1994 and 40 countries that hosted subsidiaries of US-firms. On the other hand, (Aitken and Harrison 1999) reports a negative relationship between FDI and technology growth for FDI in Venezuela. Even more confusing are the results from (Girma and Wakelin 2001), that studies inward FDI in the UK. It finds that the effect of FDI depends on the country of origin. Japanese investments yield a negative effect of FDI on technology, but US-based FDI generate a positive spillover. A comprehensive empirical study for transition countries is (Fabry 2000) with mixed results again.

The question of the transfer channel of technology left the researchers with a counter-intuitive observation. The relatively distant way to transfer technology embodied in exported goods seems to be a much better technology transmission mechanism compared to the much more intimate and close relationship that international investment establishes. Several studies attempted to use better, more fine-grained data. (Keller 2001) reviews comprehensively the empirical literature that investigates the several transfer channels. Again, the conclusions remain mixed.

Otherwise, possibly a change in perspective provides a solution to this discrepancy between intuition and empirical results. What if the inconclusive observations are not a data issue but the representation of unexplained mechanisms of knowledge transfer in international investment projects?

Taking up the aforementioned alternative perspective, in this chapter a theory will be developed that gives explanations for several empirical findings related to the question of technology transfer in multinational corporations. In particular, it makes the prediction that large differences in technology levels between the countries may prevent a technology transfer at all. In addition, for medium technology levels small technology transfers happen. But if the extent of technology transfer is at the discretion of the investor the technology transfer vanishes and a persistent technology differential sustains.

Only for medium and small technology differentials where the investor is not able to influence the knowledge spillover a technology transfer occurs.

The starting point for the theory is the observation that local participation is crucial for technology transfer. Thus, the organization of a foreign direct investment as a joint venture of a local firm and the investor plays a key role for the transfer of technology. In particular (Kokko 1994), (Blomstrom and Sjoholm 1999) and (Cassiman and Veugelers 2000) find empirical support for the idea that local participation matters.

(Keller 2001) identifies various lacks and gaps of the theory how a technology transfer might happen in an internationally diversified firm. These may be mouth-to-mouth conversation, learning from experts and other non-codified ways of knowledge spillover. This supports conceptually the importance of local participation for knowledge transfer in international investment.

To establish the theory, the very recent and quite obscure literature that deals with the often observed but unexplained phenomenon of joint venture instability is employed. This field of literature provides ideas that might explain the failure of technology transfer in international investments. In addition, it gives interesting and surprising policy recommendations how to deal with international joint ventures to optimize the technology spillover into the host country, either to increase or to prevent them, depending on the point of view.

To accomplish the task to provide a theory of technology transfer via international joint ventures and reveal the relation to the field of the joint venture life-cycle, the relevant empirical and theoretical literature is discussed first. The model itself is introduced in two steps. First, via a static approach basic properties of the model are established. In a second step, this static model is embedded in continuous time to achieve the results. Several empirical and theoretical applications, empirical predictions, and policy considerations are supplied in the concluding discussion.

3.2 Review of the literature

3.2.1 Empirical literature

Starting with the first comprehensive study on the determinants and the characteristics of international joint ventures, (Franko 1971), it became apparent that joint ventures are mostly a transitory phenomenon. Several studies affirmed instability as an inherent structural characteristics of international joint ventures. Table (3.1) gives only a small overview.

Although empirical investigations are abundant, rigorous econometric analysis of the determinants and causes of joint venture formation and breakdown is rather limited. Two studies, (Franko 1971) and (Kogut 1989), are important and often cited cornerstones with respect to an econometric exploration of the key factors in this field. They are considered herein.

3.2.1.1 Firm policy and joint venture instability (Franko 1971)

The very first extensive review of international joint ventures was accomplished by (Franko 1971) and dates back to 1971. The data set used in this book comes from a survey of 1100 US-owned international joint ventures in foreign countries in the years from 1961 to 1968. Organizational and management theories form the theoretical background for the empirical analysis.

The rich observational material revealed the high instability of joint ventures compared to other forms of international organization of the firm.

Instability of joint ventures is operationalized therein as buy-out of local partners by the investor, sell-out to local partners and liquidation of the joint venture. 28.5 % of the joint ventures in the sample ended up in one of these ways.

To identify key determinants different cross-section analysis of variance and covariance procedures are employed. 'Firm policy' changes of the partners are identified as the most important cause for breakdown.

From the point of view of modern economic analysis the theoretical equipment is insufficient. The notion of a 'firm policy', as good as may sound in the ears of business researchers, is considered today as the implicit or explicit codification of former experiences of the firm. Thus the term 'firm policy' is just a name for hidden economic mechanisms.

On the other hand, the data driven approach and extensive use of econometric methods distinguishes this work from most of its successors. One has to admit that most of them fell short compared to the technical standard set out in the book (Franko 1971). It took considerable time for an advancement of empirical research in this field.

3.2.1.2 The risk of joint venture breakdown (Kogut 1989)

The article (Kogut 1989) is mostly referred as the starting point of formal research in this area. Kogut considers data of 92 US american enterprises with respect to the joint venture decision. The material was acquired in the years from 1975 to 1983 by the publication "Mergers and Acquisition". It was collected by two mailings to 475 firms. Whenever a joint venture was proceeded a follow-up questionnaire was sent to the firm every following year.

Although the theoretical foundation is not explicitely based on a particular model the article broadly refers to the cooperative outcome of particular super-games, as for instance (Schelling 1960) and (Axelrod 1984). From these ideas Kogut derives the notion of reciprocity as the gravitational force in joint venture formation, while market competition is the centrifugal force in the joint venture partnership.

Particular interesting is the empirical implementation. It is one of the earliest econometric applications of hazard rate models. The application of a time dependent model allows to determine the risk of breakdown depending on certain covariates.

Kogut has a broader focus of joint venture breakdown than we have. Kogut considers termination of the joint venture project, dissolution, and full acquisition by one partner as instability outcomes. Due to the study design these outcomes were well observable.

The main issue provided by (Kogut 1989) is that the joint venture is both a locus of competition and reciprocity. On one hand, both firms compete, on the other they benefit from their cooperation. As long as these countervailing forces can be balanced the joint venture sustains.

The empirical implementation is done via a rank-ordered Cox hazard rate model. The concomitant variables are derived by plausibility with respect to the idea of balance between centrifugal and centripetal forces in an enterprise.

Long-term relationships between joint venture partners and common cost reducing pre-market activities, as for instance R&D, are identified as proxies for the centripetal forces. Joint ventures with close relationships between the partners and high R&D-intensity are expected to be relatively stable.

A high marketing activity, because it increases market competition, as well as a high scale intensity, because it necessitates higher market concentration, are expected to have a negative influence on joint venture stability.

These variables are used for individual firms as well as at an aggregate level to incorporate branch-level effects.

As a result, it is shown that ties between the joint venture partners and a high R&D-intensity of the joint venture reduces the breakdown risk indeed. Tighter competition and higher concentration of the industry increases the breakdown risk, as it is expected. Marketing activities and industry growth are insignificant.

With respect to the question, how technology transfer and joint venture stability are related, the study (Kogut 1989) unambiguously answers that a higher R&D activity of a joint venture increases its stability.

Article	Entities	Observations	Data	Exit	Analytical approach	IR[a]	Findings
(Franko 1971)	US-based Joint Ventures	1100	cross-section 1961-1968	buy-out, sell of, liquidation	Regressions, ANOVA	28.5%	firm policy changes cause breakdown, institutional arrangements are crucial
(Berg and Friedman 1978)	US-based Joint Ventures	123	cross section	sell-out, liquidation	Tabulation	40.7%	JV short lived
(Killing 1983)	Joint Ventures in US and Europe	35	cross section, case studies	termination	Tabulation	35%	majority JV are more stable than minority JV
(Gomes-Casseres 1987)	US subsidiaries of foreign firms	5933	Panel 1900-1975 with attrition and exit	buy-out, sell-out, liquidation	Tabulation	33%	WOS more prone to termination than JV
(Harrigan 1988)	inter-firm alliances	895	cross-section	termination	Tabulation	54.8%	strategic asymmetry increases instability
(Kogut 1989)	JV in US in manufacturing	92	survey, panel with attrition and exit	dissolution, termination, buy-out, sell-out	hazard rate model	29%	ties and high R&D intensity decreases JV termination risk, higher concentration and competition increase risk
(Desai, Foley, and Hines Jr. 2001)	US-based FDI	48025	Panel 1982-1997 with attrition and exit	buy-out, sell-out, liquidation, dissolution	Tabulations, cross section discrete choice	45.3-52.7 [b]%	Age, Ownership share at entry, Tax burden increase exit probability, profitability reduces exit probability

Table 3.1: Selected results from the empirical literature

[a]Instability Rate, i.e. share of joint ventures that exited the host country via one of the exit paths specified in column "Exit"

[b]The numbers are given for various share distributions at entry in 1989-1994

3.2.2 Theoretical literature

If one starts to review the theoretical literature on international joint ventures one might be surprised. Seldom a field of such an enormous interest in business administration and empirical research was ignored as completely as the field of multinational joint ventures, its formation, its breakdown and its life-cycle as a whole.

On the other hand, one can observe recently a raising interest in this field. If one takes into account that most of the relevant modeling techniques stem from the field of industrial organization this neglection may become understandable. The industrial organization approach to international economics is relatively recent compared to the traditional general equilibrium modeling strategy.

Moreover, a joint venture is a cooperative activity. Cooperative economic behavior is considered as an exception from the normal competitive behavior of firms in the tradition of economic theory. Models that formulate cooperative results as the outcome of competitive behavior are quite recent compared with the long tradition of theories of competition.

The theoretical offspring of the literature of the life-cycle of international joint ventures can be easily found in the theories of R&D joint ventures[1]. Although the theory of the joint venture life-cycle shares several features with the literature of R&D joint ventures, it goes one step beyond. It seeks not only an explanation of joint venture formation. It also attempts to explain the adaptive processes of a joint venture over time and the joint venture's eventual breakdown.

Several questions need to be answered by a theory of the joint venture life-cycle. First, what are the incentives for the local firm to participate, what are the incentives for the investor to share the firm with a local partner? Usually, the investor has a technology advantage. The local firms seek to participate in this advantage. Moreover, the local partner expects

[1] A particular subsection is devoted to the starting point of that literature (D'Aspremont and Jacquemin 1988). Other contributions are for example (Amir 1999). Empirical results can be found in (Siebert, Röller, and Tombak 1997).

learning or spillover from the foreign partner. The investor usually tries to acquire knowledge about the local market conditions by sharing his firm or seeks to exploit the better familiarity of local firms with the local conditions.

These incentives create the necessity to explain why neither the investor nor the local partner chose alternative cooperation agreements as licensing or inter-firm alliances.

Another important question related with the investment of a foreign firm is the appropriate formulation of the market entry problem. What are the respective market structures before and after the entry of the joint venture or the foreign enterprise? Why do barriers to entry not exist, or how these were circumvented by the investor?

The formation of the joint venture usually creates some kind of spillover that reduces the initial advantage of the investor. This spillover process requires an appropriate modeling.

The spillover changes the terms under which the joint venture was formed. How does the participation structure of the joint venture respond to these changes?

At a final stage, the reduction of the initial advantage of the investor causes a breakdown of the cooperation. What are the appropriate exit paths? The joint venture may be dissolved. A licensing agreement might replace the former joint venture. Most often, the joint venture is replaced by a wholly owned subsidiary, where the investor has bought-off the ownership share of the local partner. Nevertheless, sell-outs by the local firm can be observed, too.

The articles reviewed here form the core of an expanding literature about the joint venture life-cycle. The essential articles are considered in separate subsections. For easy reference, an overview with respect to the model setup and outcomes is provided in table (3.3).

Because this theory originates from the industrial organization literature on R&D joint ventures which incorporates external effects of technology, the offspring of that literature is considered first.

3.2.2.1 Cooperation and rivalry (d'Aspremont and Jacquemin, 1988)

The focal point for research in the field of endogenous cooperative behavior under the presence of technology spillover is the small, but highly influential paper (D'Aspremont and Jacquemin 1988). Many features of the theory developed therein are shared and extended by the theory of the joint venture life-cycle.

The main insight of (D'Aspremont and Jacquemin 1988) is that product market competition needs not to imply rival behavior in pre-market activities, as for example R&D. Moreover, cooperation can improve welfare. So these, literally, handful of pages boosted an explosive research in cooperative outcomes of rival games.

The essentials of their model building can be summarized by

- Cournot competition in product market,
- R&D activities that decrease costs as a pre-market activity,
- R&D activities of one firm decrease the R&D cost of the other firm via spillover.

Their model forms a two-stage game. On the first, the pre-market stage, R&D expenditures of the rivals are determined. The second stage is the market interaction. This is modeled as a Cournot oligopoly game with symmetric firms. The game is solved backwards by computation of sub-game perfect Nash equilibria.

In the model (D'Aspremont and Jacquemin 1988) demand and production cost are assumed to be linear in all of their arguments. R&D cost is assumed to be quadratic. Innovations, as a result of R&D expenditures, are assumed to be non-drastic[2] and purely process-oriented[3].

[2] A drastic innovation reduces the marginal cost to a level where the innovator can set the monopoly price. See (Tirole 1988), Section 10.1.3

[3] An innovation is process-oriented if it decreases marginal cost. The alternative is a product-oriented innovation that introduces new goods.

This spillover effect is modeled in the tradition of (Dasgupta and Stiglitz 1980). An investment of one firm in R&D reduces the R&D cost of the rival.

Three different outcomes can be identified as possible solutions. First, and not very surprising, no cooperation on both the R&D and the product market stages is a possible outcome. We refer to this result as the non-cooperative result. This outcome reminds of the already cited article (Dasgupta and Stiglitz 1980), where parallel research leads to excessive R&D expenditures. Cooperation can avoid the duplication of research efforts and increase the profit for both partners.

The second outcome, where the rivals decide to form a joint venture on both stages, is quite natural. R&D is accomplished in a joint venture. On the market stage the partners behave monopolistically and share the profit. This is the cooperative result.

Finally, the authors show that a third outcome is possible. Driven by the attempt to internalize the external R&D effects the firms form a joint venture on the R&D stage and are Cournot competitors on the product market stage. This result is referred to as partially competitive.

With respect to welfare, it is shown that all results were inferior to a social planner's optimum. This raises the question how the second-best outcomes can be ranked.

In ordering their results with respect to the welfare outcome it turns out that given the spillover is large enough the partially competitive outcome is welfare superior compared both to the noncooperative and the cooperative outcome.

Cooperative behavior on the R&D stage is superior to competitive behavior because duplication of research is avoided in a joint venture. Thereby external effects of R&D are perfectly internalized. This yields reduced incentives for free-riding. Thus, a higher level of research is chosen than in the non-cooperative situation. In addition, research is more efficient because research it is not duplicated.

The increased R&D expenditures in a joint venture reduces the unit costs on the product market stage. However, the Cournot competition on the product market makes the partially competitive result welfare superior compared to the cooperative solution.

So, while the partially cooperative outcome is intermediate on each stage of the game, it is superior for the game as a whole.

With the insight, that joint ventures are not an obscure phenomenon, but a legitimate outcome of *rival* behavior, developed by (D'Aspremont and Jacquemin 1988), the focus of researchers has shifted considerably towards the analysis of cooperation. So a more in-depth investigation of cooperative behavior of market rivals boosted.

For our considerations the dynamic extensions of this simple model in the towards the explanation of the whole joint venture life-cycle are crucial.

3.2.2.2 The pioneer (Kabiraj 1999)

The often discussed empirical observation that joint ventures are mostly a transitory relation between the partners did not find a response in the theoretical literature for a long time. So the authors (Yan and Zeng 1999) began their abstract with the verdict that the literature has

> " (a) lack of clear conceptualizations, inconsistent operationalizations, lack of clarity in the relationship with performance, static focus on the eventual destination rather than a developmental process, and lack of managerial relevance. "

In particular, the problem of joint venture dissolution or instability was considered as an important research issue both on theoretical and practical grounds[4].

The policy implications of joint venture formation and breakdown play an important role for developing countries, emerging markets and transition economies. These countries seek to catch up the economic strength and efficiency of industrialized countries. For this purpose joint ventures

[4]See table (3.1) for a collection of references to empirical investigations.

are considered as the most important mean of technology and knowledge transfer.

In addition, local participation constraints are induced to stay in control of the foreign capital in the host country. Thus joint ventures with local participants are the preferred mode of organization of a foreign direct investment to capture foreign know-how and technology.

Consequently, it was very disappointing that many joint ventures were withdrawn, dissolved or bought off. To investigate the reasons and consequences of joint venture breakdown a stronger theoretical foundation was required.

The first theoretical work that explicitely deals both with joint venture formation and joint venture breakdown in a dynamic, unified way is the article (Kabiraj 1999). This paper overcomes the critique of a static environment, gives clear conceptualizations of joint venture breakdown and establishes the link between performance and evolution of the joint venture over time.

The model setup has some prototypical features. A technologically advanced investor enters the market in a technically less developed foreign country. Two firms with identical cost are already established on the foreign market. All firms have costs that consist of two parts, marketing and distribution cost and production cost. Although the indigenous firms lack technological capabilities, thus have a higher production cost than the investor, they have a cost advantage in distribution and marketing.

The firms engage in Cournot competition in every period repeatedly. The investor has three choice options. The first is to enter the foreign market and act as an additional competitor. In this case the profit is the outcome of a three-party Cournot oligopoly. This is a consequence of the assumption that cost advantages are strictly symmetric. The sum of all costs is equal for all firms, accordingly. If the investor can not access the local knowledge about marketing and distribution no cost advantage exists.

Besides this, the investor can make a cross-border licensing arrangement with a local firm where each partner cross-licenses his specific cost advantage to the partner. In this case the market participants play a Cournot game where two firms have lower costs than the third.

The last option is to form a joint venture with one of the local firms. For a joint venture to be established, the participants must incur a fixed cost. In this case the firms proceed to compete in quantities, but the costs of the joint venture are lower than the costs of the local firm without a foreign partner. In addition, the Cournot game is played by two partners only, because the investor and one local firm have merged.

It is shown that depending on cost advantages and fixed costs several equilibria exist where pure Cournot competition, cross-licensing and joint venture are the dominant strategies, respectively. Moreover, for the cross-licensing and the joint venture case it happens that the respective non-participant vanishes from the market.

The entry of the investor into the foreign market via joint venture allows the outside firm to innovate by imitation at a fixed cost. This leads to a reduction of the production cost down to the level of the investor. If this happens it becomes more profitable for the investor to dissolve the joint venture friendly and to leave the technology for the firms on the foreign market. This is the result of the increased competition. Competition has increased by the innovation, because the cost advantage of the joint venture has vanished.

If the joint venture is dissolved the indigenous firms and the investor in the host country's market engage in Cournot competition again, but at lower cost now. Without the possibility of imitation the joint venture would have had lasted forever.

The 'spillover' concept used is best described as 'spillover by external pressure'. Of course, this is not a true spillover in the sense that external knowledge is acquired directly. However, at an aggregate level technology catch-up is correlated to joint venture formation for such an innovation model, too[5].

[5]The idea of innovation as a result of the entry of foreign investors into a local market was introduced independently in (Bertschek 1995) for an oligopoly model, which concluded that market entry of advanced firms indeed increases the incentive to innovate. (Bertschek and Lechner 1998) tested this theory with

The central issue of (Kabiraj 1999) is the decision of a technologically advanced investor to acquire a part of a local firm versus cross-licensing of mutual advantageous knowledge. In other words, it is for the host country the question to make an innovation, if a joint venture is formed, or to buy it from outside via cross-licensing.

The occurrence of local imitation results in a breakdown of the joint venture. Thus this model explains the whole joint venture life-cycle indeed. It endogenizes both the formation and breakdown of the joint venture.

The breakdown concept used herein is friendly dissolution with the exit path of a licensing agreement. From an empirical point of view a friendly dissolution of a joint venture into a licensing relationship is seldom found. Most joint ventures are dissolved by buy-out or unfriendly dissolution. Another, possibly even more grave objection is the observation that licensing contracts sometimes preceed joint venture agreements with the same firm[6].

Within the framework of a 'make or buy' decision for the host country with respect to innovations a model of the innovation process is incorporated in (Kabiraj 1999). A fixed payment is required to innovate. The higher the payment the faster is the innovation. With this extension it follows that the joint venture would always break down because innovation is sure. But the optimal date of innovation has to be determined.

The author gives the optimal timing of the innovation and, within the given context, of the joint venture dissolution. One finds that the larger the initial technological advantage of the investor the faster the joint venture is resolved. This is surprising outcome. It states that long-term joint ventures are connected to low technology, while high technology joint ventures break down fast. The economic reason is the increased incentive to imitate for the firm outside the joint venture if the technology advantage is large.

To summarize, (Kabiraj 1999) provides the first formal model which explains joint venture formation and joint venture breakdown in a dynamic,

---

data successfully.

[6]The dataset (Universität München 2002) provides anecdotical evidence for the media industry, chemical consumer products, and in general for products where market reputation (brands) play an important role.

unified model. It is introduced in the context of a 'make or buy' decision for the host country with respect to the advanced technology of the investor, where a joint venture, the 'make' decision for the host country, or a licensing agreement, the 'buy' decision for the host country, are the options available for the investor.

Joint venture formation is caused by the exploitation of mutual synergies. Joint venture breakdown is the outcome of imitative innovation by an outsider, who undermines the technology advantage of the joint venture. Innovation in turn results from increased incentives for the outsider firm due to entry of a foreign investor. The observation of aggregate spillover in this model would be described best by the concept of 'spillover by external pressure', that is, increased competition increases the incentive to innovate.

The breakdown concept used is the friendly dissolution of a joint venture into a licensing agreement. The empirical importance of such an exit concept is negligible.

The most interesting result is that a large technology gap between the joint venture and the local firm reduces the lifespan of the joint venture more than a small technology advantage.

3.2.2.3 Local participation requirements and the effect of governmental policy (Sinha 2001)

Another article that deals with the problem of endogenous joint venture formation and breakdown and its relation to technological imitation is (Sinha 2001). Again, the topic is the question of joint venture formation versus licensing. In this model the joint venture formation is the result of an exogenous local participation requirement. Joint venture instability arises from a, possibly stepwise, reduction of this requirement by the host country's government and the threat of imitative innovation by local firms.

The assumption of an exogenous local participation requirement is not appealing in a model which seeks to explain the joint venture life-cycle endogenously. On the other hand, many developing countries imposed local participation requirements on the activities of foreign firms.

With the multinational agreement on investments (MAI) such restrictions may be sanctioned now by the World Trade Organization. Consequently, it is of considerable interest to explore the consequences of liberalization with respect to ownership restrictions of foreign direct investment projects.

The model of (Sinha 2001) has three players. Two players, the firms, are able to act strategically. The third player, the government, can only make announcements with respect to ownership liberalization. It is not able to act strategically.

The model forms a two-period multi-stage game. In the first period, the host country's government sets a minimum participation constraint for a joint venture in favor of a local firm. The second period starts with the governmental announcement to abandon this requirement, either partially of fully.

In the first period, after the governmental restriction, the investor offers the local firm his advanced technology for a share of the local firm and a fixed up-front fee. If this offer is not accepted all players get a payoff of zero. The investor remains at home and the indigenous firm continues to supply the home market with inferior technology as a monopoly.

If the offer is accepted a joint venture is formed. The technology advantage of the investor is assumed to be drastic. It follows, that the joint venture can operate as a monopoly. As a borderline case, the local partner could pay the investor without giving away any share. This results in a license agreement where the investor transfers his technology without possession of a share of the local firm. One may note, that without a local participation requirement a joint venture would have never occurred. If the initial offer is rejected the game ends.

In the second period, the ownership restriction laid upon the joint venture is announced to be reduced or abandoned at all. Now the investor offers the local partner to buy an additional share of the local partner or to sell out a part of his ownership share. In addition, with some positive probability the local partner achieves a drastic imitative innovation, yielding a decrease of marginal costs. As a consequence, the local firm could act as a monopoly now. If the indigenous firm fails to innovate the advanced technology has to be acquired from the investor.

The outcomes of the second period depend on the success of the imitation, the initial local participation share, and the governmental policy. The various possibilities are summarized in table (3.2).

	Partial Liberalization	Full Liberalization
Imitation Success	stable or adjustment in favor of local firm	stable or adjustment in favor of the investor
Imitation Failure	instable, adjustment in favor of investor	instable, adjustment in favor of investor

Table 3.2: Outcomes for different government policies and R&D results of the local firm

The adjustments in favor of the investor, if local imitation has failed, is quite naturally the result of the increased bargaining power of the investor. And, because the participation constraint is liberalized, the threat of the investor to withdraw the investment and set up a wholly owned subsidiary becomes more credible. The increased credibility of the threat of the investor to abandon the joint venture is the reason that under full liberalization the participation share could be adjusted in favor of the investor, irrespective of imitation success.

Solving backwards from the final stage, that is summarized in table (3.2), the following results can be derived. If for partial liberalization a joint venture occurs in the first period it is continued at first period's conditions if imitative innovation was successful, it is continued with a readjustment in favor of the investor if innovation failed. If the joint venture did not occur in the first period the investor offers the most advanced technology in period two via licensing. In this case the investor captures all positive profits and the local firm makes zero profits.

If for full liberalization a joint venture occurs the investor could offer continuation of the contract if imitation was successful or readjust in his favor depending on the parameters. If the imitation failed the investor always readjusts in his favor.

In general terms, full liberalization weakens the position of the local firm more than a stepwise liberalization does, because full liberalization increases the credibility of the investor to withdraw his investment.

The model introduced by (Sinha 2001) provides an explanation for joint venture formation which is exogenous only. Without governmental measures the investor would always set up a wholly owned subsidiary driving the local firm out of the market. In contrast, joint venture breakdown is explained endogenously as the result of local imitation. However, it is a quite strong assumption that firms can repeatedly innovate in large steps as the repeated drastic innovation suggests.

A critical point is that the only exit channel that actually occurs is the exit via licensing of the technology by the investor, similar to (Kabiraj 1999). Although dissolution is always a threat by the investor, it is always weakly dominated by a licensing agreement.

Although the exogeneity of joint venture formation is not satisfying, the interactions between governmental measures and local participation responds to a lack in the previous literature. Hence, it provides hints for a liberalization policy with respect to joint ventures.

A particular attractive feature is the rich dynamics which the model provides. It generates adjustments in favor of the investor, as well as in favor of the local partner, as well as stable joint ventures. An integrated model of the joint venture life cycle should be able to explain all of these three empirically observable possibilities.

3.2.2.4 Mutual synergies (Roy Chowdhury and Roy Chowdhury 2001)

Among the sparse literature considering the dynamics of joint venture participation under the presence of learning or spillover the work (Roy Chowdhury and Roy Chowdhury 2001) is up to date the most comprehensive approach. It provides for the first time a welfare analysis of joint venture breakdown. Similar to the models discussed above, (Roy Chowdhury and Roy Chowdhury 2001) employs an game-theoretic approach. And again, a technologically superior investor and a local partner join to exploit synergistic advantages in the market of the host country.

The incentive for the local firm to participate in the joint venture is the chance to enjoy the technology advantage of the investor and, possibly, to learn the foreign technology. For the investor the incentive is the knowledge of indigenous firms about the local conditions. In this model these are the peculiarities of the host country with respect to marketing and distribution.

The model considers two periods. In each period the game is played in two stages. The solution concept is the sub-game perfect Nash equilibrium.

On the first stage in every period, the investor and a local partner sequentially decide to participate in a monopolistic joint venture or to pursue Cournot competition as a duopoly.

The investor makes the first draw. If the investor decides to participate in a joint venture the local partner can agree or disagree. If the local partner agrees a joint venture is set up or proceeded, respectively. If one partner disagrees the joint venture is not formed or dissolved.

Once the organizational form is chosen, the agents decide the output level. If a duopoly arises, i.e. no joint venture was formed or proceeded, the parties simultaneously decide their output level on the foreign market. If a joint venture is formed the parties decide on their respective input levels for the joint venture production.

It is assumed, that neither the input levels nor the output level are contractible *ex ante*. The only contractible characteristics is the sharing rule for the profits of the joint venture. It is assumed to be equal sharing. The authors legitimate this assumption with a bargaining argument but do not qualify this further.

Production is described by a standard production function with the inputs labor and capital. It is supposed, that both parties have identical production functions, but the investor has lower capital cost than the local partner while the local partner has lower labor cost. Similar to the model in (Sinha 2001) the cost advantage is assumed to be perfectly symmetric, that is, with respect to the overall cost neither the investor nor the local firm have an advantage over each other. Formally, with r_i, $i = Investor, LocalPartner$ as capital cost and w_i, $i = Investor, LocalPartner$ as labor cost, it is assumed that $w_{Investor} > w_{LocalPartner}$, $r_{LocalPartner} > r_{Investor}$ and $w_{LocalPartner} = r_{Investor}$. Under this assumption the investor supplies capital and the local partner supplies labor.

If both parties form a joint venture in the first period both partners can reduce their respective cost disadvantage by mutual learning. With respect to learning symmetry is assumed again. Hence the cost-reducing effect is the same for both parties.

The breakdown concept used herein is unfriendly dissolution. The game can be solved backwards, starting in period two on stage two. Altogether the solution of the game can be completely summarized by the ordering of the three possible profit outcomes.

First, one has to consider the per-period joint venture payoff per firm $\overline{J}$. Second, one has the payoff $\tilde{\pi}$ in a situation where a joint venture was formed in the first period which breaks down in the second period yielding Cournot competition. Third and last, one has to consider the case of two-period Cournot competition which yields the per-period payoff $\hat{P}$. The sequence of Cournot competition in the first period followed by a joint venture in the second period needs not to be considered because the technological parameters remain stable. Consequently, Cournot competition is never followed by joint venture formation.

If the ordering of outcomes is $\overline{J} \geq \tilde{\pi} > \hat{P}$ then, irrespective of the outcome in the first period, in the second period a joint venture arises. Thus, one needs to consider the first period only. But because $\tilde{\pi} > \hat{P}$, that is, a first period joint venture followed by Cournot competition dominates the two-period Cournot payoff, the joint venture arises in the first period, too. This is the stable joint venture outcome.

If the ordering of outcomes is $\tilde{\pi} \geq \overline{J} > \hat{P}$ then one has the case of a joint venture formation in period one and a breakdown in period two.

The last case to be considered, is where $\tilde{\pi} > \hat{P} > \overline{J}$. This constellation implies Cournot competition in the second period irrespective of the first period. The formation of a joint venture in period 1 depends on the parameters of the game. With a discount factor δ a first-period joint venture arises if the discounted profits from a first period profit followed by Cournot competition with lower cost dominates the two-period Cournot payoff. If this is the case, one has the case of joint venture breakdown again. Otherwise, Cournot competition prevails in every period.

Comparative statics unveils the role of learning in this model. It is found that increased learning increases the incentives to form a joint venture. This comes from the rise of the second period's profit. But if learning is too fast, it increases the probability of a second period breakdown also. This is a result of the non-contractibility of inputs. It is more profitable to exploit the own knowledge than to rely on the relatively small cost advantage of a partner that possibly engages in free-riding.

For short, increased learning encourages initial joint venture formation and final joint venture breakdown.

Note, that without the additional ingredient of non-contractability of inputs or output, the assumptions of the model would yield a persistent joint venture from the first period on. This would be the Pareto superior solution, too[7].

This is the result of the fact, that with contractible inputs, it is always (weakly) preferable to employ the mutual advantages as a whole instead of imperfectly acquired knowledge.

The most distinguishing part of the article is the welfare analysis. It states that joint venture formation may be advantageous, but joint venture breakdown, if it occurs, is always Pareto superior.

This sharply contrasts to the economic policy folklore[8] that joint ven-

[7]One has to be careful with a general verdict on this topic in a partial model, of course.

[8]See for example (Zschiedrich and Keller 1999)

tures should be persistent to be a success. In contrary, in the light of the model the opposite is true.

So we can summarize

- if input or output would be contractible joint venture formation would always occur,
- learning encourages joint venture formation in the first period and joint venture breakdown in the second period,
- joint venture breakdown is Pareto dominant.

But of course there are some points that should be criticized. Joint venture breakdown in this model is the result of the non-contractibility of inputs or output. Although this is not a critical point *per se*, it is highly implausible in the model context. If both firms have identical and perfectly symmetric production functions which are common knowledge, it is questionable that an informed strategic agent is not able to infer from his own behavior into that of the other rational agent.

A second point that makes the employed model a quite special case is the assumption of perfect symmetry of the firms. The consequences of asymmetries are not discussed. Such a special case makes generalizations of the insights complicated.

And a last important point is, that the sharing rule is exogenously given. The interest of the current research is just the role of ownership as an incentive device. With the exogeneity of the distribution rule this important question becomes intractable.

Nevertheless, the comparative statics section and the welfare discussion clearly distinguish the article (Roy Chowdhury and Roy Chowdhury 2001) from the predecessors discussed so far. The Pareto superiority of a joint venture breakdown in the context of the model should renew the discussion on the role of joint ventures beyond pure managerial issues.

Article	Market Structure		Innovation	Spillover[a]	Joint Venture Characteristics					Breakdown[b]
	pre-formation	post-formation			formation	breakdown	exit[c]	partial[d]	dynamics[e]	
(Kabiraj 1999)	symmetric duopoly	3-oligopoly	non-drastic, process	external	endogenous	endogenous	licensing	no	investor	friendly dissolution
(Sinha 2001)	symmetric duopoly	monopoly	drastic, process	internal	exogenous	endogenous	licensing	yes	both	buy-out, sell-out
(Roy Chowdhury and Roy Chowdhury 2001)	monopoly	monopoly	non-drastic, process	internal	endogenous	endogenous	dissolution	no	investor	unfriendly dissolution
this thesis	competition, monopoly, Cournot oligopoly	competition, monopoly, Cournot oligopoly	non-drastic, process	internal and external	endogenous	endogenous	buy-out, sell-out	yes	both	buy-out, sell-out

Table 3.3: Overview theoretical literature

[a] Does the spillover occur within the joint venture (internal) or does it affect other firms (external)?
[b] Breakdown Concept
[c] Exit path
[d] Is partial adaption of shares possible?
[e] Can the investor adapt the share, can the local partner adapt the share, or both?

3.2.3 Summary and conclusion

We can identify the key concepts for a theory of joint venture that views joint ventures as a transitory dynamic phenomenon of international economic activities of firms.

The fundamental setup starts with a technologically advanced investor that enters a foreign market. The technology advantage usually is embodied in lower production cost. Local firms possess country specific advantages, usually on the labor market. Because both firms have a specific advantage, both have an incentive to join. The investor can participate in the knowledge of the local firm about the market conditions in the host country. For the local firm the incentive is twofold. First, it can enjoy the foreign technology in a joint venture. Second, it is expected that the local firm can learn from the foreign technology. The effect from technology participation is an internal effect that applies to the joint venture partners only. The learning effect is an externality that even outsiders may benefit from. And finally, learning or spillover is simultaneously the source of both technological catch-up and joint venture breakdown.

Contrary to the public good character of technology in the 'new' macroeconomic growth theory[9], herein technology transfer requires interaction between a knowledgeable party and an ignorant party to be transferred. In other words, technology is an infectious[10] 'disease' instead of a freely accessible, ubiquitous well of health.

The literature reviewed so far bears some complications. An important problem is the proper definition of the market structure. All models considered used oligopolistic or even monopolistic market structures. This requires the specification the strategic options of the economic agents. The strategic decision problem may hide the role of economic fundamentals in the formation and breakdown of joint ventures as well as their implication

[9]See for example (Romer 1986), (Lucas Jr. 1988), or, again, the well-known monograph (Grossman and Helpman 1991)

[10]There is a literature that indeed applies models of epidemics to the diffusion of technology. See (Azariadis and Drazen 1990) for a theoretical approach and (Chong and Zanforlin 1999) for an empirical application.

for technology transfer.

As the investor can always enter the foreign market with a wholly owned subsidiary[11], one has to consider a market entry effect besides the technology and learning effect. The entry of the investor in the local market increases the competition, thus it drives down prices. The learning or spillover effect drives prices in the same direction. For an economic analysis of the effect of a joint venture a researcher needs to isolate these effects and to assess the relative importance of the two. Under imperfect competition this procedure is not perfectly valid because the market structure effect implies the learning effect and vice versa. It is, consequently, logically impossible to separate one effect from the other.

Empirically, joint ventures are not necessarily bound to a particular market structure. It may be of interest to develop a model that allows joint venture formation and breakdown irrespective of a particular market structure. Such a model would allow to separate a market entry effect from the effect of technology spillover.

The modeling of the learning or the spillover effect is another topic of relevance in these models. An appropriate approach should specify how and how much technology is transferred. All models considered in the review specified the amount of knowledge that is transferred as an exogenously given jump. The extent of knowledge transfer should be endogenized.

All local firms have the ability to jump via learning to the level of technology the foreign firm supplies. If this were the case the empirical research in this field would not have encountered the difficulties to identify foreign direct investment as a technology transfer channel. One might suspect that technology is transferred in small portions instead of one large jump. It might be useful to find a modeling specification that allows a stepwise and possibly incomplete transfer of knowledge which is endogenously specified. Such a model might explain the mixed evidence of technology transfer in FDI.

Closely related to the idea of stepwise technology transfer is the ne-

[11]Of course, with the notable exception of (Sinha 2001), who excluded this option by assumption.

cessity to model a stepwise adaption of the share owned by the local firm. Again, from empirical reasons, a stepwise adjustment of shares seems more attractive than an immediate exit from the joint venture. It may give the partner the chance to adapt the participation in the shared firm depending on previously transferred knowledge. Moreover, it is a reasonable idea that the participation share may influence the amount of knowledge transferred. Usually, the access to know-how of a partner in the enterprise is the larger, the larger the share held by the partner. On the other hand, the idea of knowledge as a 'contagious' element may justify the idea that some portions of knowledge are accessible to a local partner without the approval of the investor.

Finally, the appropriate exit path for the joint venture participants needs careful consideration. Although there are several possibilities, the exit path from joint venture to a licensing agreement as chosen in (Kabiraj 1999) and (Sinha 2001) is the option that has the weakest empirical support. In contrary, there are examples where a licensing agreement preceeded the formation of a joint venture.

From an empirical point of view, the termination of the joint venture and a buy-out are the most frequently observed exit paths[12]. This should be taken into consideration for a meaningful theory of the joint venture life-cycle.

The investigation of the key determinants of joint venture breakdown is sparse compared to the vast empirical material. The study (Franko 1971) identified organizational issues as the key determinants. But the economic content beyond business administration issues remained vague.

In contrast, the study (Kogut 1989) was able to identify competitive elements and the R&D-intensity of the joint venture as the main determinants for the breakdown risk of a joint venture. This econometric diagnosis establishes the empirical link between technology transfer and the joint venture life-cycle.

---

[12]See table (3.1)

3.3 Optimal ownership in international joint ventures and technology transfer–the static case

3.3.1 The economic environment

Let us suppose, there is an investment opportunity in a foreign country. An investor from the home country wants to engage in the host country to exploit this possibility.

The foreign market is perfectly competitive. As the investor has an advanced technology, embodied in marginal cost lower than the marginal cost of the host country, the local firms seek to form a joint venture with the investor to participate in the cost advantage. However, the cost advantage is not drastic.

Forming a joint venture with a local partner gives the local firm insights into the technological and managerial skills of the investor. This provides the local partner the chance to learn from the investor. The learning process creates a spillover that reduces the technological advantage of the investor. The investor and the local partner may wish to adapt their respective ownership shares accordingly. This process is essentially dynamic. Thus, it is appropriate to study the evolution of the ownership distribution over time as a dynamic, continuous process.

If local participation reduces the profit prospect of the investor abroad, why should the investor agree with local participation? The local government does not interfere with the market, hence, no local participation requirements are levied on the investor. Contrary to the previous chapter, the government does not even tax the firms.

The incentive for the investor to share his investment project is the ability of the local firm to reduce the investor's cost even further. Under the appropriate conditions a joint venture may form.

This setting reduces the modeling structure of joint venture formation and, possibly, breakdown to the economic fundamentals. In this environment strategic options are limited to the choice of the ownership share.

This approach has several advantages compared with the traditional multi-stage game-theoretic modeling. First, the inherent strategic timing problem for participants in the market can be neglected. The investor and the indigenous partner chose their strategies in the very beginning and because they are always on their equilibrium path, there is no incentive to deviate afterwards. Competition limits the available options to the adaption of the ownership share.

Second, due to the simplification of the strategic decision process, the specification of the information structure becomes very simple. Every parameter can be observed or derived perfectly from the observations. With the appropriate parameterization of the model it can be shown that perfect competition acts as a revelation mechanism. The only assumption necessary is the non-contractability of the effort of the local partner. Even this assumption serves convenience. It can be shown that the non-contractability can be derived from the model setup. However, because this treatise focuses on the properties of a technologically static joint venture versus a joint venture with a changing technology gap over time, further discussion of the topic of contractability deviates too far from the main line of argumentation.

And last, the reduction of strategic choices unveils the role of economic fundamentals in the decision process on the ownership of the investment project, as there are, for instance, demand elasticities and cost.

The study of the dynamic behavior of international joint ventures is proceeded in three stages. First we study the optimization problem of the investor in a static environment. No spillover arises in this case. Note, that this case can be easily extended to a dynamic situation without spillover. Then we discuss the dynamic approach with spillover that is independent of the actions of the economic agents. Finally, the decision problem of the investor is studied when spillover depends on the magnitude of local participation. Several empirical predictions, applications, and policy recommendations of the theory developed herein finish this chapter.

Before the model is specified in detail, it is necessary to explicate the necessary concepts.

3.3.2 Definitions

Usually, a foreign direct investment project is denoted a joint venture, whenever an investor founds newly an enterprise with a partner. If the enterprise has already existed, but the ownership structure has changed, it is considered as a 'merger and acquisition'. With respect to the discussion of technology transfer between the investor and a local partner this distinction is irrelevant.

In accordance with the preceding literature on international joint ventures the concept of an international joint venture used herein is broader than the corresponding concept in business research. The central issue is that the ownership in a foreign direct investment project is shared with a local partner. The novelty of the investment project does not play any role.

Definition 3.1 *A foreign direct investment project is a joint venture if a foreign investor holds a share of a firm in the host country of the investment, but does not own the firm as a whole.*

In addition, it is necessary to describe joint venture stability and instability more precisely.

Definition 3.2 *A joint venture is said to be j-stable if for all* $t \in [0, T]$

$$\frac{s_j(T)}{\sum_{\substack{i=1 \\ j \neq i}}^{M(T)} s_i(T)} = \frac{s_j(t)}{\sum_{\substack{i=1 \\ j \neq i}}^{M(t)} s_i(t)},$$

where $M(0) > 1$ *and* $M(t)$ *denotes the number of members of the joint venture at a time* t *and* s_i *is the share of the joint venture partner* i, $s \in [0, 1]$ *Otherwise it is called j-instable.*

One might ask if this formulation is not a superfluous one. Suppose, the investor I permanently changes his second and only partner P, keeping the own share constant over time. Would such a joint venture called to be stable? With the above definition this joint venture is I-stable, but never P-stable.

The other important problem to care for is the entry of new partners into a joint venture. As long as the relative share of the partners with respect to all others is kept, this joint venture is stable with respect to every partner already participating. It is only instable with respect to the new partner. So the above definition, provides a formal base for the analysis of the redistribution of the ownership shares in joint ventures.

An additional advantage is that one can discuss the issue of stability of ownership with respect to whole groups of owners. Possibly a group of owners holds its share constant as a group, but within this group the shares are redistributed. In such a case it may suffice to consider the stability with respect to this group only.

The simplest case of a joint venture arises if we have only two partners or groups of partners. Let us denote one group with I for the investor and the other one with L for the local partner.

Then the definition becomes

Definition 3.3 *Suppose we have a joint venture with two partners, the investor and the local partner, with their respective shares* $1 - s$ *and* s, $s \in (0, 1)$. *The joint venture is called to be instable if for some points in time* t *and* T $s(t) \neq s(T)$. *It is denoted to be stable otherwise.*

In case of instability the investor has made a buy-out if for $t < T$ *follows that* $s(t) - s(T) > 0$. *The investor has made a sell-out if for* $t < T$ *follows that* $s(t) - s(T) < 0$.

Now we are left with some borderline cases. What happens when all partners end up with a share of 0? Then the joint venture is terminated. This case is not considered herein.

The case with one partner holding 100% is not considered as a joint venture in general. Nevertheless, it is sometimes appropriate to consider this situation as a borderline case of a joint venture. It will be explicitely mentioned if necessary.

As the review of the literature revealed, the concept of joint venture breakdown may be understood in different ways. The breakdown concepts used in this treatise are buy-out and sell-out. They are collected in a separate definition.

Definition 3.4 *A joint venture breaks down by buy-out if the investor acquires a share of 100 % of the joint venture. A joint venture breaks down by sell-out if the local partner acquires 100 % of the joint venture. A joint venture breaks down by dissolution if both partners withdraw their participation in the joint venture. The enterprise ceases.*

3.3.3 The static model of international joint ventures

The investor engaging in a foreign country has an instantaneous, quadratic cost function

$$C^I(q) = \frac{1}{2}q^2 + q. \tag{3.1}$$

The q variable denotes the quantity produced. Note, that we have for the cost function $C'^I > 0$, $C''^I > 0$[13], that is, the cost is convex in quantity.

The cost function for the local firms is assumed to be

$$C^L(q) = \frac{1}{2}q^2 + (1 + A)q \tag{3.2}$$

with A being a technical parameter describing the technological lack of the local firms compared to the investor. It follows, that the investor has lower marginal cost as long as $A > 0$. Trivially, with $A = 0$ the marginal cost of the investor and the firms of the host country are equal. Consequently, $A \geq 0$ is assumed. Again, we assume for the cost function that $C'^L > 0$, $C''^L > 0$.

It is assumed, that there are no fixed costs, and that there is perfect foresight for every economic agent in the market. Free entry forms a competitive fringe for the firms in the market. Under these assumptions,

[13] The specific form chosen for the cost function is a tribute to later proceedings, namely for the dynamic optimization problem with spillover.

the marginal cost function of the local firms, $MC^L(q)$, coincides with the inverse supply function, $S^{-1}(p)$, by virtue of profit maximization.

The free entry condition in connection with the quadratic cost function yields that there exist N firms that produce a positive quantity and gain positive profits[14]. The aggregate supply would be N times the individual supply, $S(p)$. As usual, the supply function $S(p)$ is a mapping from the non-negative price space into the quantity space. Because entry is not considered in this model, N can be normalized to one without loss of generality, following the argumentation in (Das 1987). Thus, the aggregate supply can be identified with the supply of a single firm.

If the investor forms a joint venture with a local partner the investor can benefit from the market specific knowledge of the local firm. This is incorporated in the instantaneous cost function (3.1) via the parameter $\theta \in [0, 1]$[15] that reduces the marginal cost of the investor. Thus, we have for the cost function of the joint venture

$$C^J(q, \theta) = \frac{1}{2}q^2 + (1 - \theta)q. \tag{3.3}$$

The cost reduction requires an effort by the local partner. It is captured by an effort function $\psi(\theta)$ with $\psi' > 0$, $\psi'' > 0 \ \forall \theta > 0$, and $\psi'(0) = \psi(0) = 0$. These assumptions for the effort function imply that without appropriate remuneration the indigenous firm makes no effort at all for cost reduction of the investor. In addition, the effort is assumed to be non-contractible.

It is important to note, that this formulation entails a fundamental asymmetry between the local firms and the investor. While local firms can not benefit from a joint venture among themselves, a joint venture between the investor and a local firm creates synergies.

This assumption is not unreasonable. For instance the local firm can possess knowledge about the quality of workers on the local market. For a joint venture with other local firms this might be useless knowledge,

[14]For a more comprehensive discussion of the relationship between cost function and the long-run market structure see for example (Varian 1992), Chapter 13 or (Mas-Collel, Whinston, and Green 1995), Chapter 10.F.

[15]The notation $[a, b]$ denotes all x of the real line where $a \leq x \leq b$ as usual.

because it does not create a cost advantage. In connection with the technological advantage of the investor, however, this knowledge might create additional value in a joint venture. This asymmetry is a purely synergetic effect.

One may criticize this assumption. However, the alternative, that a mutual cost reduction is perfectly symmetric, appears to be even more unrealistic.

Now the demand side of the local market is considered. To avoid technical complications the class of all possible demand correspondences is restricted to a particular set of linear demand functions. The necessary properties of the demand functions are collected for convenience in a definition.

Definition 3.5 *Let $1 + A$ denote the reservation price of supply with $A \geq 0$. Let $\frac{g}{h}$ denote the reservation price of demand. Then a linear demand function $q(p) = g - hp$, that is a mapping from the price domain $P = [0, \infty)$, into the quantity domain $Q = [0, \infty)$, is called regular if for an $A \geq 0$ each of the following conditions holds:*

1. $h > 0$
2. $\frac{g}{h} > 1 + A$
3. $g > 2h + h^2$.

This definition should be considered as a convenience device instead of a full characterization of demand functions for which the further investigations are valid. Some remarks may give an intuitive idea why the conditions are necessary and how they influence the properties of the model. Properties (1) and (2) of the definition (3.5) guarantee a strictly monotonically declining demand curve for which the reservation price $\frac{g}{h}$ is higher than the reservation price of supply, $1+A$. In other words, one can expect a competitive equilibrium for a fixed technology advantage A with an equilibrium price above $1 + A$.

However, this is not sufficient to guarantee the compatibility of a non-drastic technology advantage, A, with a competitive equilibrium because the marginal cost can be influenced by the cost reduction parameter θ, too. Condition (3) of definition (3.5) ensures this compatibility as we will see below.

Now the ingredients of our model are complete. Two problems need to be adressed to ensure that the problem is stated properly. First, does there exist for the technology advantage A exist a nonempty intervall to vary? Otherwise a consideration of the question with a varying technology advantage would be useless.

Second, if such equilibrium set exists, what is the appropriate pricing rule for the investor? Of course, both problems are related. The problem of price setting depends on the set of equilibria determined by the intersection of the demand and supply curve. On the other hand, the set of possible equilibria depends on the prices the investor can set.

First, the question of existence is answered[16].

Lemma 3.1 *Let $q(p)$ be a regular demand function, and let the cost functions be defined by the formulae (3.1) and (3.2). Let the price p^m_θ denote the monopoly price of the investor if the effort θ is incurred and q^m_θ the respective quantity. Similarly, the price p^L_A denotes the competitive price on the local market if the technology gap is A and q^L_A denotes the respective quantity. Then there exists an interval $\mathcal{A} = [0; \overline{A}]$ so that for every $\theta \in [0, 1]$ and every $A \in \mathcal{A}$ the monopoly price is larger than the local competitive price, that is*

$$p^m_\theta > p^L_A. \tag{3.4}$$

Proof: Because the monopoly price for $\theta = 1$ and $A = 0$ is the lowest possible monopoly price it suffices to show that

$$p^m_{\theta=1} > p^L_{A=0}. \tag{3.5}$$

[16] I am grateful to Sven Rady who pointed out a severe error in a previous version and provided the appropriate corrections.

To this end, we show that $p^m_{\theta=1} > p^L_{A=0}$ is valid. From the strict monotonicity of the linear demand function we can conclude that $p^m_\theta > p^L_A$ implies $q^m_\theta < q^L_A$. Moreover, this is true for $p^m_{\theta=1}$ and $p^L_{A=0}$ and their respective quantities, too. This in turn is equivalent to the condition that

$$\frac{g}{2+h} < \frac{g-h}{1+h}. \tag{3.6}$$

It is sufficient that $g > h^2 + 2h$ for condition (3.6) to hold. This was assumed by the condition (3) of definition (3.5) of the regularity of the demand function.

The existence of a whole interval of for which the inequality (3.6) is valid can be concluded from the continuity and the monotonicity of the demand and the supply function and the fact that the already proven inequality (3.5) is strict. ***q.e.d.***

Again, it is emphasized that this lemma guarantees that we have a whole interval for the technological parameter A to vary, where both competitive equilibria exist and the investor can not set the monopoly price. Otherwise a consideration of a dynamic model where spillover may change the A would be useless. But consistence of non-drastic technology advantage with competitive equilibria was shown so far for constant marginal costs only[17]. The lemma (3.1) extends these results.

The argumentation is collected in figure (3.1). The local market equilibrium is the point E where demand $q(p)$ and supply MC_L cross.

The price $p^m_{\theta=1}$ determined by the point M, which is the intersection of the marginal revenue curve MR and the investor's supply MC^I where $\theta = 1$ and $A = 0$, is the lowest upper bound for all possible competitive prices. That $p^m_{\theta=1}$ is the lowest possible monopoly price follows from the property that for strictly decreasing demand functions the monopoly price is a non-decreasing function of the marginal cost[18].

Lemma (3.1) argues that even for the lowest monopoly price an interval for competitive equilibria with prices below $p^m_{\theta=1}$ exists. Accordingly, such

[17] See (Tirole 1988), Chapter 10.1

[18] See (Tirole 1988), pp. 66-67

an interval exists even for higher marginal cost of the investor. The line MC_A may be such a marginal cost curve that would provide a competitive equilibrium with $p^m_{\theta=1} > p^L_{A>0}$.

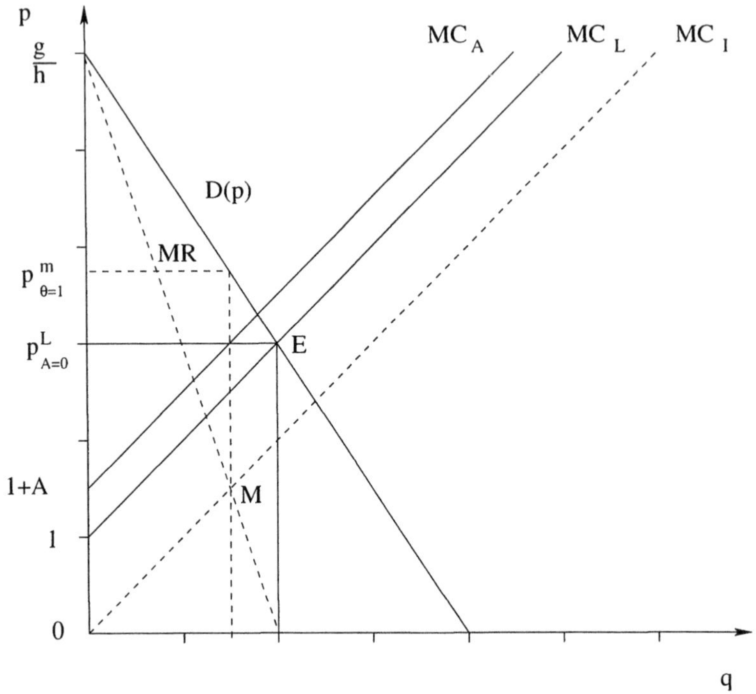

Figure 3.1: Graphical exposition for lemma (3.1)

Now, as the existence of an appropriate equilibrium is ensured, we can state the pricing rule of the investor. For convenient reference, it is cloaked into a lemma:

Lemma 3.2 *For every $A \in [0, \overline{A}]$, the investor sets his price p^I equal to the equilibrium price p^L of the local market.*

$$p^I = p^L \tag{3.7}$$

Proof: Suppose otherwise, there should exist an $A \geq 0$ where $p^I \neq p^L$. The investor's price can not be above it because it is evidently not profit maximizing.

Thus the investor could only set the price $p_L - \varepsilon$ where $\varepsilon > 0$. But this can not be an equilibrium because the profit of the investor is increased with $p_L - \frac{\varepsilon}{2}$.

Thus at the margin (with $\varepsilon \to 0$), the investor sets the price p^L. Again, this must hold for all $0 \leq A \leq \overline{A}$. ***q.e.d.***

The consequence of lemma (3.2) is that only the local market equilibrium is relevant for the price and quantity setting. These in turn depend on the technology gap A. So the profit does.

For the further proceedings some notation and explanations are required.

Definition 3.6 *Let $p(A)$ and $q(A)$ denote the equilibrium price and quantity for the technology advantage $A \geq 0$. Define the profit function of the wholly owned subsidiary as*

$$\pi = \pi(A) = p(A)q(A) - \frac{q^2(A)}{2} - (1-\theta)q(A). \tag{3.8}$$

It is a mapping from the technology domain $A = [0, \overline{A}]$ into the real line: $\pi : A \to I\!R$.

Accordingly the optimization problem of the investor can be formulated as follows.

$$\max_{s \in [0,1]} (1-s)[\pi(A) + \theta q(A)] \tag{3.9}$$

$$\text{s.t.} \quad \theta \in \operatorname*{argmax}_{\theta \in [0,1]} \{s[\pi(A) + \theta q(A)] - \psi(\theta)\}. \tag{3.10}$$

From the condition (3.10) one can immediately conclude that the local partner incurs a positive effort only if he is remunerated for it. Or to state it formally, $\psi(\theta) > 0$ if and only if $s > 0$. Because the effort function is strictly monotonic one gets that $\theta > 0$ if and only if $s > 0$.

For the solution process we need to replace the constraint (3.10) with a more handy condition. For this purpose we use the first order condition[19]. So we get a modified version:

$$\max_{s\in[0,1]} (1-s)[\pi(A)+\theta q(A)] \tag{3.11}$$

$$\text{s.t.} \quad \psi'(\theta) = sq(A). \tag{3.12}$$

Again, the later discussion will be very burdensome without a particular specification of the effort function. So we simply set

$$\psi(\theta) = \frac{1}{2}\theta^2 \tag{3.13}$$

and get another formulation of the above problem

$$\max_{s\in[0,1]} (1-s)[\pi(A)+\theta q(A)] \tag{3.14}$$

$$\text{s.t.} \quad \theta = sq(A). \tag{3.15}$$

This version is considerably simpler but entails the essential features of the problem of interest.

In a final step the participation constraint of the local partner (3.15) can be incorporated into the profit function of the investor which reduces the investor's optimization problem further to as single line

$$\max_{s\in[0,1]} (1-s)[\pi(A)+sq^2(A)]. \tag{3.16}$$

The replacement of θ by the first order expression $sq(A)$ justifies the following notational shorthand

Definition 3.7 *The function*

$$\pi_s = \pi_s(A) = (1-s)(\pi(A)+sq^2(A)) \tag{3.17}$$

denotes the profit function of the investor for a joint venture with the local participation share $s \in (0,1]$.

[19]The validity of this approach is proved in appendix B.1

3.3.3.1 Model solution and comparative statics

The solution for the investor's optimization problem is straightforward.

Proposition 3.1 *The maximization problem has the following solution for the participation share of the local firm*

$$s^* = \begin{cases} 0, & if\ \frac{3}{4} + \frac{1-p(A)}{2q(A)} \leq 0 \\ \frac{3}{4} + \frac{1-p(A)}{2q(A)}, & if\ \frac{3}{4} + \frac{1-p(A)}{2q(A)} \in (0,1) \\ 1, & otherwise. \end{cases} \tag{3.18}$$

Proof: Using the Kuhn-Tucker conditions under consideration of the corner conditions. ***q.e.d.***

An example might illustrate the discussion so far.

Example 1 *Now let us assume that the demand is linear Take a regular demand function*

$$q(p) = g - hp.$$

For the marginal cost curve we have

$$p = q + 1 + A.$$

Thus the equilibrium prices and quantities are computed as

$$p(A) = \frac{g+1+A}{1+h} \tag{3.19}$$

and

$$q(A) = q(p(A)) = \frac{g-h-hA}{1+h} \tag{3.20}$$

respectively.

With $g = 5$ and $h = 1$ and $A = 0$ one finds the equilibrium price of $p = 3$ and the demanded quantity as $q = 2$. One easily checks that this parameterization generates a regular demand function.

Using the formula (3.18) we find that $s = \frac{1}{4}$ is the optimal share for the local partner. This situation is illustrated in figure (3.2). It shows

functions π_s for varying s and A fixed at zero. For the equilibrium price $p = 3$ the respective curve $\pi_{\frac{1}{4}}$ is above the functions π_s for all other s, in particular it is located above $\pi_{s=0} = \pi$
The profit computation gives a profit of $\pi_0 = 2$ in this case if the investor keeps the investment project for himself. Sharing the investment project gives the investor a profit of $\pi_s = \frac{9}{4}$.

The local partner is paid for his effort level of $1/2$ gets from his joint venture share a payment of $3/4$.

So we find that local participation increases the profits from 2 to 3 that is an improvement of 100%.

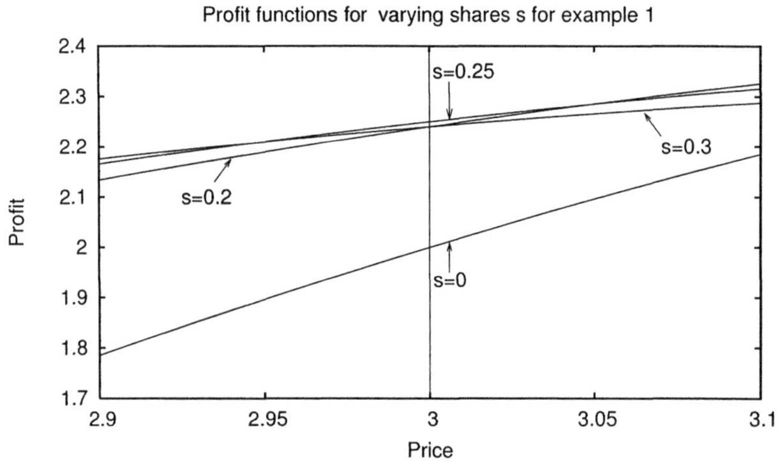

Figure 3.2: Profit functions for various s for example 1

Now, as the optimal participation share is determined we can turn to the comparative statics results. First, the relationship between the demand elasticity and local participation is considered.

Proposition 3.2 *The higher the price elasticity of demand the smaller the participation share s for $s > 0$.*

Proof: From the definition of the price elasticity ε of demand we find for a fixed A that

$$q = -\frac{dq}{dp}\frac{p}{\varepsilon}.$$

Thus, with formula (3.18) we get

$$s = \frac{3}{4} - \frac{p-1}{p}\frac{\varepsilon}{2h}. \tag{3.21}$$

Due to the regularity of the demand function we find that $p - 1 > 0$ and $h > 0$. Differentiation with respect to ε yields a negative derivative, accordingly. ***q.e.d.***

The economic idea implied by this result can be easily explained. A small elasticity means that small change in the price yields only a small change in the quantity requested by consumers. In this case it is more useful for the investor to reduce the cost than to reduce the price. Otherwise, a high price elasticity makes it less attractive for the investor to reduce the cost. A slight reduction of the price is more effective to increase the profit instead.

Note that the equation (3.21) gives an interesting additional insight into the decision calculus of the investor. The term $\frac{p-1}{p}$ is the Learner index for marginal costs of one. The Learner index in turn gives the markup of the equilibrium price over the marginal cost. In our case it is the markup of the equilibrium price over the unit cost of 1. Thus one could say that the optimum share is a proportion of the markup of the equilibrium price over the unity unit cost.

For the question of technology transfer the relationship between technological handicap A and the local participation share s is of particular interest. This relationship will establish a leitmotif for the rest of this treatise. We find the following relationship.

Proposition 3.3 *The ownership share s the investor gives away is a monotonically decreasing function of the technological advantage A for $s > 0$.*

Proof: The optimal share is given by equation (3.18). Differentiation with respect to A gives

$$\frac{d}{dA}s = p_A \left(-\frac{1}{2q(A)} + \frac{1-p(A)}{2q^2(A)}h \right). \tag{3.22}$$

Because a rising technology gap implies a higher price, the technology derivative of the equilibrium price, p_A, is positive. The expression within the parentheses is negative. To see this the terms are considered separately. The first term, $-\frac{1}{2q(A)}$, is negative because $q(A) > 0$ in equilibrium by regularity of the demand function. The second term $\frac{1-p(A)}{2q^2(A)}h$, is negative because $h > 0$ and $1 - p(A) < 0$ by regularity of demand. It follows that a declining technology advantage increases the incentive for the investor to give away a larger share of his investment project. ***q.e.d.***

The economic argumentation is straightforward. The cost reduction due to local participation becomes the more important the smaller the technology advantage of the investor is. So we get a tendency that investors with a smaller technological potential more easily give up some share of their investment.

An important consequence of the comparative statics results is that there must exist an upper bound A^* for all A beyond that local participation never happens. This is the situation in formula (3.18) where $\frac{3}{4} + \frac{1-p(A)}{2q(A)} \leq 0$.

Example 1 continued: *Now let us compute this situation for the example. First, we check the comparative statics result. The local participation share should be higher if A decreases and vice versa.*
Let us set $A = 1$. Then, as expected the equilibrium price is higher $p = \frac{7}{2}$, consequently $q = \frac{3}{2}$. Using formula (3.18) for the different A we find that $s = -\frac{1}{12}$. Thus $s^* = 0$. Indeed with rising A the local ownership share decreases.
One more step will determine the 'reservation' advantage A^* for local participation. Again, formula (3.18) will be employed. Setting s=0, we get that $A^* = \frac{g-h}{2+3h} = \frac{4}{5} = 0.8$. If the technology advantage exceeds 0.8, that is, the unit cost of local firms are 80% higher than those of the investor,

the investor will keep his investment project for himself. He enters with a wholly owned subsidiary.

The inverse relationship between technology advantage and participation share of the local partner creates a puzzle. The argument delivered by proposition (3.3) is perfectly intuitive. Why then, when a technologically advanced investor enters a market by joint venture, one can observe that in most cases joint ventures break up in a buy-out? The answer to this puzzle can be found in the dynamic properties spillover create. The introduction of spillover in section 3.4 will turn this static outcome upside down.

Another empirically relevant conclusion of the comparative statics result of proposition (3.3) is that the number of minority owned joint ventures should be higher with a higher technological similarity of the home and the host country.

One might these results attribute to the particular market structure used. However, as the appendix B.2 demonstrates the results with respect to local participation remain valid for a monopolistic investor. This provides an argument that the settings of the model remain valid for various more complicated market structures.

So let us conclude. First, the more similar countries are with respect to technology the higher the probability that the investor enters the foreign country via joint venture.

Second, the more similar the countries are, the higher is the share the local partner gets.

Third, there exists a maximum level of the technology differential above which firms enter as a wholly owned subsidiary only. These investment projects remain a wholly owned subsidiary forever.

3.4 Optimal ownership in international joint ventures and technology transfer– the dynamic case

3.4.1 Introduction

A meaningful dynamic element in the context of the model of international joint ventures developed so far is introduced by a change of the technology advantage over time. A simple time integral would do nothing more than just multiply the static solution with some constant. But a change in the technology parameter A introduces new behavioral degrees of freedom for both the investor and the local firms.

A change in technology is the result of technology diffusion from the advanced partner to the handicapped partner. Thus, one can hope that the technology levels of the local firm and the investor converge over time. The question arises, if the investor is able to control the flow of information to the local partner or not. The ability to control the flow of knowledge may influence the behavior of the joint venture partners considerably. Hence, it is necessary to distinguish between autonomous and controllable spillover[20].

An autonomous spillover occurs if the investor is not able to influence the spillover, except for a buy-out. A controlled spillover means that the knowledge flow can be supervised and controlled by the investor. It does not mean that it can be totally prevented, again, with the exception that a joint venture is not formed at all.

The following considerations can be summarized in two general results. First, a technology transfer in a joint venture occurs if and only if it is instable. And more interesting, a joint venture breakdown by buy-out occurs if and only if the spillover is controllable.

These results link the recent debate about the technology transfer mode

[20](Blomstrom and Kokko 1996) introduced a similar concept. They distinguished between passive and active spillover. The passive spillover coincides with autonomous spillover if applied to joint ventures, but an active spillover requires a costly action of the partner who seeks to acquire knowledge.

in multinational corporations and foreign direct investment[21] with the literature on joint venture instability.

The discussion of dynamic models of the joint venture life-cycle is started with the results of autonomous spillover. With autonomous spillover the investor's share declines over time until the technology advantage is fully dissipated. Consequently, autonomous spillover explain the phenomenon of a rising share of the indigenous partner and, as an extreme case only, a sell-out.

The introduction of controllable spillover reverses this result. It is shown that a joint venture with controllable spillover breaks down by a buy-out. It is emphasized, that both the investor and the local partner agree with respect to this exit path. The intuitive explanation for this counter-intuitive behavior of the local firm is that the spillover not only undermines the profit prospect of the investor. Because in a joint venture the profit is shared among the partners, the spillover effect reduces the profits of the local partner as well.

The introduction of controllable spillover considers the adaptive process of a microeconomic model in continuous time. Because the approach via optimal control theory has some complications the approach via the calculus of variations is chosen. Instead of a direct solution for the shares $s(t)$ as a control variable, the optimal technology path $A(t)$ will be derived. Using this solution the optimal share path can be recovered.

The investigation is finished with a discussion of empirical predictions, policy implications, and the application of the theory to the topics of absorptive capacities and the mixed evidence of technology transfer in foreign direct investment.

3.4.2 Notation

As a prologue to the model, the already established notation is extended to the time dimension.

[21] See for a review of the very extensive discussion, for example, (Keller 2001) or (Blomstrom and Kokko 1996).

Definition 3.8 *The technology advantage becomes a function, $A(t)$. The initial technology advantage is denoted as $A(0) = A_0$, similarly the final technology advantage at the moment T is denoted as $A_T = A(T)$. The time derivative of technology is denoted as $\dot{A}$, the second time derivative as $\ddot{A}$.*

The notation $A > 0$ is a shorthand for the expression 'for all t where $A_t > 0$'. This applies to the expressions for s, q and p accordingly.

The functions of A, $p(A)$, $q(A)$ and $\pi(A)$ become functions of time, too. They are denoted as p_t, q_t, and $\pi(t)$, respectively.

Because q depends on time only via the price, which in turn depends on time via $A(t)$, we need to consider the instantaneous price derivative of q. It is denoted as $\frac{dq}{dp}$, and it is time dependent, too. Thus, it is $\frac{dq}{dp} = \frac{dq}{dp}(t)$ for a particular moment t in time.

Similarly, a notation for the technology derivative of the price is required. It is denoted as p_A as a shorthand for $\frac{dp}{dA}(t)$.

For the profit of the investor, $\pi(t)$, the time derivative is $\dot{\pi}$. Similarly to the static model we will see that the notation π_s for the profit of the investor when the joint venture is shared applies to the dynamic case, too. Accordingly, the definition of π_s can be used in dynamic context as $\pi_s(t)$. The time derivative is denoted as $\dot{\pi}_s$. To avoid misunderstandings the notation $\pi(t)$ and $\pi_s(t)$ is used for the time dependent profit exclusively.

The effort and the share become dynamic, too. For the effort the notation θ_t is used. For the ownership share $s(t)$ and s_t is used interchangeably. However, s_0 and s_T denote the initial local participation share and the final local participation share. The time derivative of the local participation share is $\dot{s}$.

3.4.3 Ownership dynamics with autonomous spillover

We start our discussion with the case of autonomous spillover. The simplest specification arises with a constant spillover. The cost differential is reduced in a fixed time interval for a fixed amount. The spillover sustains as long as the technology advantage is positive, and the local firm takes part in the investment project. Technology diffusion stops, if the technology advantage is lost, or if the joint venture is dissolved.

Formally, the autonomous spillover is defined by

$$\dot{A} = \begin{cases} -\beta, \text{ with } \beta > 0 & \text{if } A > 0 \text{ and } s > 0 \\ 0 & \text{otherwise .} \end{cases} \tag{3.23}$$

As simple as it is, we will see that this specification already captures the central feature of the dynamic model with autonomous spillover. We are interested in the relationship between the time path of shares, $s(t)$, and the time path of the technology gap, $A(t)$. The intuition from proposition (3.3) lets us expect a rise in the share of the local partner as the technology advantage shrinks.

The optimization problem of the investor is

$$\max_{s_t \in [0,1]} \int_0^{\infty} e^{-\rho t}(1 - s_t)[\pi(t) + \theta_t q_t] dt \tag{3.24}$$

$$\text{s.t.} \quad \dot{A} = \begin{cases} -\beta, \beta > 0 & \text{for } A > 0 \text{ and } s > 0 \\ 0 & \text{otherwise} \end{cases} \tag{3.25}$$

$$\text{s.t.} \quad \theta_t \in \tag{3.26}$$

$$\operatorname{argmax}_{\theta_t \in [0,1]} \int_0^{\infty} e^{-\rho t} \left\{ s_t[\pi(t) + \theta_t q_t] - \psi(\theta_t) \right\} dt$$

Because the condition (3.25) is independent of the ownership share $s(t)$, we can solve the differential equation for $A(t)$ independently from the optimization problem of the investor. Doing so, we get the relation

$$A(t) = A_0 - \beta t \tag{3.27}$$

where A_0 is the initial technology advantage.

As in the static case, the optimization constraint can be replaced by a first order condition. Differentiation with respect to θ_t reveals that the first order condition for the local partner in the dynamic model is identical to the first order condition in the static model. Altogether, these replacements yield

$$\max_{s_t \in [0,1]} \int_0^\infty e^{-\rho t}(1 - s_t)[\pi(t) + \theta_t q_t]dt \tag{3.28}$$

$$\text{s.t. } A(t) = A_0 - \beta t \tag{3.29}$$

$$\text{s.t. } \theta_t = s_t q_t. \tag{3.30}$$

With this formulation one finds that the optimization problem of the investor does not affect the path of technology. As a consequence, it can be treated as an instantaneous problem to be solved at every moment of time.

The independence of the side condition (3.29) of the optimizer s_t delivers a formula for the time path, s_t, that formally looks like the condition in the static case (see formula 3.18), but now it describes a function of time t:

$$s_t = \frac{3}{4} + \frac{1 - p_t}{2q_t}. \tag{3.31}$$

From this result, we can derive the inter-temporal behavior of the local participation share.

Proposition 3.4 *For regular demand, $q = g - hp$, and autonomous spillover according to (3.25) the local participation share the rises over time:*

$$\dot{s} > 0. \tag{3.32}$$

Proof: We take equation (3.31) and compute the time derivative. This yields

$$\dot{s} = -p_A \dot{A} \left(\frac{1}{2q_t} + \frac{1 - p_t}{2q_t^2} \frac{dq}{dp} \right). \tag{3.33}$$

The expression $p_A \dot{A}$ is negative, because the price decreases with a decreasing technology gap, and $\dot{A}$ is negative by assumption. The expression in parenthesis is positive. To see this, we need to remember that a regular demand function implies equilibrium prices above $1 + A_t$ for all relevant

A_t. This is a result of lemma (3.1). Thus, $1 - p_t < 0$. The price derivative of demand is negative by assumption, too. Altogether, we get a positive quantity for this term. Thus the expression in parenthesis is positive, the expression outside the parenthesis is positive by the minus sign, too. This delivers the result. ***q.e.d.***

Proposition (3.4) makes the link to the comparative statics result of proposition (3.3) obvious. The local participation share rises over time, as expected. Of course, the joint venture is instable in this case. If it breaks down finally depends on the parameters of the problem. A sell-out is the only possible exit path in this case.

The graphic example of figures (3.3) and (3.4) illustrates the time paths of the price p_t, quantity q_t, technology advantage $A(t)$ and the local participation share $s(t)$. The parameters of the linear demand example (1) are used. The initial technology advantage is set to $A_0 = 0.75$.

Using equilibrium expressions for the prices and quantities from the example (1) yields for the investor's optimization problem (3.24)-(3.26) the optimal time path of the local participation share as

$$s_t = \frac{3}{4} - \frac{g - h - \beta t + A_0}{2(g - h + h\beta t - A_0)}. \tag{3.34}$$

According to proposition (3.4) this is an increasing function over time. One finds that the technology advantage is exhausted after $\frac{A_0}{\beta} = 7.5$ periods with the constant spillover assumption. The quantity and price at the final time T are the figures for $A_T = 0$ from example (1). Note, that all time paths are linear, except for the share path. The curvature of the share path depends on complex relations between demand and demand derivatives. These relations do not bear an economic content. To summarize, if spillover is autonomous, a decreasing technology advantage is accompanied by an increasing participation share of the local firm as long as the technology advantage exists. If the technology advantage is exhausted, the local participation share remains at the same level forever.

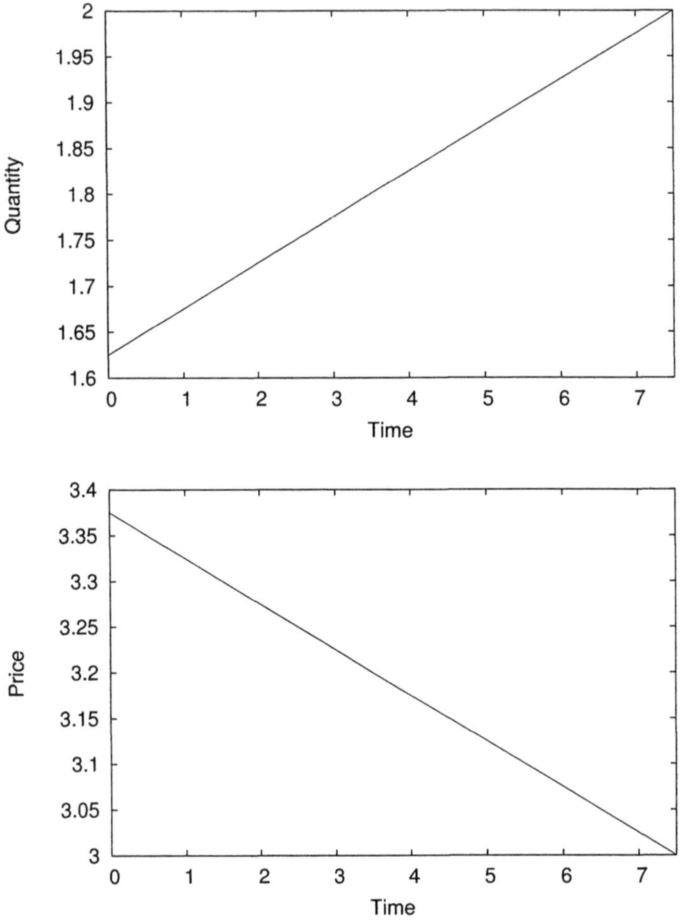

Figure 3.3: Price and quantity time paths with linear demand $q(p) = 5 - p$, constant autonomous spillover $\beta = 0.1$, and initial technology advantage $A_0 = \frac{3}{4}$

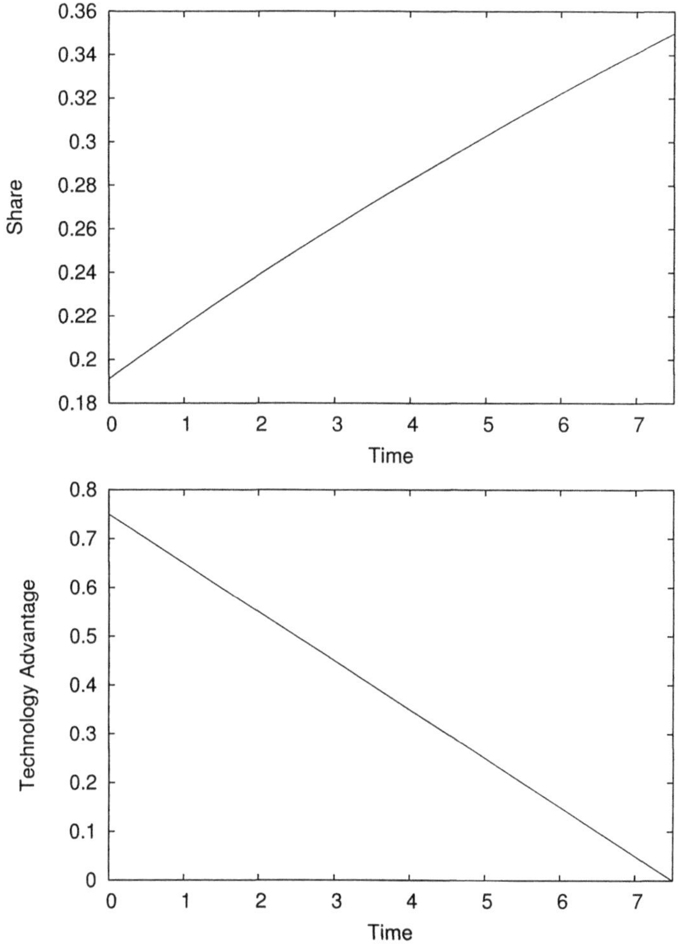

Figure 3.4: Technology and share time path with linear demand $q(p) = 5 - p$, constant autonomous spillover $\beta = 0.1$, and initial technology advantage $A_0 = \frac{3}{4}$

3.4.4 Ownership dynamics with controllable spillover

3.4.4.1 Introduction

Controllable spillover is the opposite of autonomous spillover. While an autonomous spillover appears whenever a local firm participates in the investment project of the foreign firm and can not be controlled by the participating parties, controllable spillover is subject to the discretion of the investor.

This implies, that a new decision problem arises for the investor. Under autonomous spillover the decision problem for the investor was basically the static one, applied to every moment in time. In contrast, the controllable spillover offers new degrees of freedom. The investor has to control the spillover in an optimal way at every moment in time, taking into consideration future losses due to the learning effect of the local partner.

Again, the investor has to take into account the participation constraint of the local partner. Using a first order approach it becomes the already known condition for the static problem, but now it must hold for every moment in time.

3.4.4.2 The indirect Euler equation approach

This section introduces spillover as a function of the shares the local partner owns. A particular appealing specification is

$$\dot{A} = -\beta s_t q_t \tag{3.35}$$

where β is a positive parameter, s_t denotes the ownership share, and q_t denotes the equilibrium demand. The spillover specification (3.35) states that the learning effect of the local partner is a proportion of the quantity $s_t q_t$ produced by the local partner for the joint venture. The parameter β denotes the ability of the indigenous firm to absorb the potential spillover. In addition, without a technological advantage there is nothing to learn. Formally, $A = 0$ implies $\dot{A} = 0$. Accordingly, without local participation the local firms can not learn from the investor. That is, $s = 0$ implies $\dot{A} = 0$.

The optimization problem of the investor is

$$\max_{s_t \in [0,1]} \int_0^\infty e^{-\rho t}(1 - s_t)[\pi(t) + \theta_t q_t]dt \tag{3.36}$$

$$\text{s.t. } \dot{A} = \begin{cases} -\beta s_t q_t & \text{if } A > 0 \text{ and } s > 0 \\ 0 & \text{otherwise} \end{cases} \tag{3.37}$$

$$\text{s.t. } \theta_t \in \operatorname*{argmax}_{\theta_t \in [0,1]} \int_0^\infty e^{-\rho t} \{s_t[\pi(t) + \theta_t q_t] - \psi(\theta_t)\} dt \tag{3.38}$$

Similar to the approach of the static model we replace the maximization condition of the partner with the respective first order condition. With the specific functional form of the effort function, $\psi = \frac{1}{2}\theta^2$, and by virtue of lemma (C.2), we get:

$$\max_{s_t \in [0,1]} \int_0^\infty e^{-\rho t}(1 - s_t)[\pi(t) + \theta_t q_t]dt \tag{3.39}$$

$$\text{s.t. } \dot{A} = \begin{cases} -\beta s_t q_t & \text{if } A > 0 \text{ and } s > 0 \\ 0 & \text{otherwise} \end{cases} \tag{3.40}$$

$$\text{s.t. } \theta_t = s_t q_t \tag{3.41}$$

The transformations of (3.36)-(3.38) lead to a control problem with the control variable s_t and the state variable $A(t)$.

As a solution method the side conditions of the problem (3.39)-(3.41) are incorporated into the objective function. This results in a problem of the classical calculus of variations. With this replacement, the technology gap A and its respective time derivative $\dot{A}$ become the dynamic variables the problem has to be solved for. The Euler equation for this approach turns out to be an inhomogeneous second order linear differential equation that is accessible to standard solution methods with the time path of technology as the solving function.

Once the time path for A_t is found, the respective path for the participation share of the local firm can be recovered by the spillover specification (3.35).

Of course, this indirect solution method comes at a price. Using the indirect Euler equation approach one assumes implicitly that as long as $A > 0$ the solution path has at least two non-vanishing time derivatives[22]. However, from an economic point of view such a smoothness condition is an acceptable assumption. In other words, a knowledge spillover does not jump. A spillover is a continuous flow of knowledge.

This assumption implies via (3.35) that the share path has at least one continuous time derivative for all points t in time where $s_t > 0$.

From Equation (3.35) one gets

$$s_t = -\frac{\alpha \dot{A}}{q_t} \tag{3.42}$$

with $\alpha = \frac{1}{\beta}$. Using the indirect Euler equation approach the side conditions can be incorporated into the objective functional. This yields

$$\max_{s_t \in [0,1]} \int_0^\infty e^{-\rho t} \left(1 + \frac{\alpha \dot{A}}{q_t}\right) \left(\pi_t - \alpha \dot{A} q_t\right) dt \tag{3.43}$$

with the initial technology advantage A_0 and the requirement that $A_t \geq 0$, for all t. Of course, if $A = 0$ we can apply the static theory as developed in the previous sections.

The general form of the integrand can be found to be

$$F(A, \dot{A}, t) = e^{-\rho t} G(A, \dot{A}).$$

Hence, we can apply lemma (C.3) for the derivation of the necessary optimality condition. To emphasize that the direct control variable s was replaced by the respective terms in $\dot{A}$ and A, we replace the profit function in prices and shares, $\pi(p, s)$, with its counterpart in A and $\dot{A}$, $G(A, \dot{A})$, formally,

$$\pi(p_t, s_t) = G(A_t, \dot{A}).$$

[22]See (Chiang 1992), chapter 2

This replacement increases the readability of the respective derivatives considerably.

The terms for the necessary optimality condition, according to lemma (C.3), are

$$
\begin{aligned}
G_A &= p_A(q_t + \alpha\dot{A}) + (p_t - 1)\frac{dq}{dp}p_A \\
&\quad -q_t\frac{dq}{dp}p_A - \frac{3\alpha}{2}\frac{dq}{dp}p_A\dot{A} && (3.44)\\
G_{\dot{A}} &= \alpha(p_t - 1) - \frac{3\alpha}{2}q_t - 2\alpha^2\dot{A} && (3.45)\\
\frac{d}{dt}G_{\dot{A}} &= \alpha p_A\dot{A} - \frac{3\alpha}{2}\frac{dq}{dp}p_A\dot{A} - 2\alpha^2\ddot{A}. && (3.46)
\end{aligned}
$$

Altogether, the terms (3.44)-(3.46) yield

$$
\begin{aligned}
G_A - \frac{d}{dt}G_{\dot{A}} + \rho G_{\dot{A}} &= p_A q_t + (p_t - 1)\frac{dq}{dp}p_A - q_t\frac{dq}{dp}p_A && (3.47)\\
&\quad + 2\alpha^2\ddot{A} + \rho\alpha(p_t - 1) - \frac{3\rho\alpha}{2}q_t - 2\rho\alpha^2\dot{A}.
\end{aligned}
$$

The optimal technology path can derived now.

3.4.4.3 On the optimal technology path

To get an analytic solution the regular demand function $q(p) = g - hp$ is plugged into formula (3.47). We already have the expressions for the competitive equilibrium price and quantity from the example (1). Using these, we get for the Euler equation (3.47) that

$$\begin{aligned} & G_A - \frac{d}{dt}G_{\dot{A}} + \rho G_{\dot{A}} = \\ & \frac{g-h-hA}{(1+h)^2} - \frac{gh-h^2+hA}{(1+h)^2} + \frac{gh-h^2-h^2A}{(1+h)^2} + 2\alpha\ddot{A} \\ & +\rho\alpha\frac{g-h+A}{1+h} + \frac{3\rho\alpha}{2}\frac{g-h-hA}{1+h} - 2\rho\alpha^2\dot{A}. \end{aligned} \tag{3.48}$$

We can start collecting terms to achieve

$$\begin{aligned} & G_A - \frac{d}{dt}G_{\dot{A}} + \rho G_{\dot{A}} = \\ & \frac{g-h}{(1+h)^2} - \rho\alpha\frac{g-h}{1+h} + \left[\rho\alpha\frac{2+3h}{2(1+h)} - \frac{2h+h^2}{(1+h)^2}\right] A \\ & -2\rho\alpha^2\dot{A} + 2\alpha^2\ddot{A}. \end{aligned} \tag{3.49}$$

This is a non-homogenous, second-order linear differential equation. We want to get the normalized form

$$\ddot{A} + a_1\dot{A} + a_2 A = b. \tag{3.50}$$

The solution of a differential equation has the general form $A_t = A_p + A_c(t)$. The particular integral, A_p, is any solution of the equation (3.50). The function $A_c(t)$ is the complementary integral and describes the evolution of the solving path of Equation (3.50) over time. From the theory of differential equations[23] it can be found that the complementary integral must have the general form

$$A_c = B_1 e^{r_1 t} + B_2 e^{r_2 t} \tag{3.51}$$

where B_1 and B_2 are some constant coefficients to be fixed later.

[23] see (Stepanov 1956)

The values of the characteristic roots r_1, r_2 are the solution of the characteristic equation of (3.49) and derived from

$$r_{1/2} = -\frac{a_1}{2} \pm \sqrt{\left(\frac{a_1}{2}\right)^2 - a_2}. \tag{3.52}$$

One has to find the appropriate expressions for a_1, a_2 and b. These are explicated in the following lemma.

Lemma 3.3 *From the Euler equation (3.49) the respective values for a_1, a_2 and b are derived as*

$$a_1 = -\rho \tag{3.53}$$

$$a_2 = \frac{1}{2\alpha^2}\left[\rho\alpha\frac{2+3h}{2(1+h)} - \frac{2h+h^2}{(1+h)^2}\right] \tag{3.54}$$

$$b = \frac{1}{2\alpha^2}\left[\frac{g-h}{(1+h)^2} - \rho\alpha\frac{g-h}{1+h}\right]. \tag{3.55}$$

So we are left to determine the coefficients B_1 and B_2. To this end the start and end point of the technology path would deliver the appropriate conditions. For the initial technological advantage $A(0) = A_0$ one could chose any appropriate positive number as long as it guarantees that $s(0) > 0$. Otherwise the investor would keep the investment project as a whole and no spillover would occur.

However, this delivers one equation for the two coefficients only. A second equation can be derived from the final condition of the time path of technology. However, what is the final condition? Can we state that $A(T) = A_T > 0$ or not? Is the final condition even a finite one or is it merely a transversality condition that has to be fulfilled as t goes to infinity?

To answer these questions, it is useful to investigate the economic intuition behind the general mathematical terms described so far.

The general solution of the complementary integral (3.51) constitutes a weighted average of two different time paths, $e^{r_i t}$, $i = 1, 2$, with the weights B_i, $i = 1, 2$. For an optimal time path one should expect a decline of the technology advantage with a declining rate.

Why one should expect such behavior? The declining path of the technology advantage derives from the spillover. If at the moment of entry the investor gives away a share $s > 0$, the local market can learn from the investor. Consequently, A should decline. However, as we know from the static case, proposition (3.3), this gives an incentive to the investor to give away even more shares.

Formally, s is a function of the time derivative of the technological advantage, but with the opposite sign. A declining path of A results in an increasing share for the local partner. Otherwise, giving away more and more shares of the investment project can not be an inter-temporal optimizing path because the erosion of the technological advantage yields an erosion in profits. Consequently, there should exist a moment τ where the profits with $s > 0$ are lower than the profits with $s = 0$. Hence, the investor will buy back some share of his firm on an optimal time path, because the cost gains from cooperation with a local firm are not enough to reward the investor for the losses by the declining technological advantage. This requires a decreasing rate of technology loss.

Moreover, the local partner agrees with the investor to reduce his participation share. That the local partner agrees stems from the fact that a declining technology advantage reduces the profits the local partner can get from participation in the joint venture as well. This creates an incentive for the local partner to return some shares to the investor.

Indeed, to check the behavior of the roots r_1, r_2 one plugs in the results of lemma (3.3) into the determining equation for the characteristic roots (3.52). Of course, only real roots bear economic sense. Thus, the expression under the root should be positive. This is guaranteed if a_2 is negative. The term a_2 in turn is negative if and only if

$$\rho\alpha < (1 - \frac{3}{2}\rho\alpha)h^2 + (2 - \frac{5}{2}\rho\alpha)h. \tag{3.56}$$

In general, condition (3.56) is not necessarily fulfilled. Plausibility suggests, that the real discount rate ρ should be small compared to the value of h, where h is the price derivative of demand. The values for α, that is, the inverse of the learning ability, β, should be not too large. If the local firm is able to learn from the whole quantity produced, sD, $\beta = \alpha$ is one. For example, with $\rho = 0.05$, $\alpha = 1$, and $h = 1$ one gets from (3.56) that the condition holds easily. On the left of the inequality, one has $\rho = 0.05$. On the right, we have $3 - \frac{40}{200} = 2.8$. Thus for reasonable parameter constellations, the condition (3.56) can be expected to hold. If it does, it generates one positive and one negative root, leading to a technology loss with a decreasing rate, as the economic reasoning suggested.

Again, from the general theory of differential equations we get that the particular integral is of the form $A_p = \frac{b}{a_2}$. Thus we have:

Lemma 3.4 *The particular solution of the Euler equation (3.48) is*

$$A_p = -\frac{\rho\alpha(2+3h)(1+h) + 2h(h-2)}{2(g-h)(\rho\alpha(1+h)+1)}. \tag{3.57}$$

Because the particular solution is negative, at least one of the coefficients B_i must be positive. Otherwise the technology path would be always negative. This result would be senseless within the context of the model.

A formal derivation may emphasize this intuition. In the beginning the investor has the initial technology advantage A_0. Supposed we would have some A_T as a result of $\lim_{t\to\infty} A_t$. Then coefficients B_i $i = 1, 2$ are determined via

$$A_0 = A_p + B_1 + B_2 \tag{3.58}$$
$$A_T = A_p + B_1 e^{r_1 T} + B_2 e^{r_2 T}. \tag{3.59}$$

This is a system of nonlinear equations. Solved simultaneously, these relations give the general expressions for the coefficients

$$B_1 = \frac{A_p - A_T + (A_0 - A_p)e^{r_2 T}}{e^{r_2 T} - e^{r_1 T}} \tag{3.60}$$
$$B_2 = \frac{A_T - A_p + (A_0 - A_p)e^{r_1 T}}{e^{r_2 T} - e^{r_1 T}}. \tag{3.61}$$

These equations would be well-determined if one could find an expression for the final time T or alternatively one would take the time limit to fix these expressions.

Moreover, the final condition determines if the joint venture breaks down by buy-out or if it remains in stable state forever. Of course, the latter case appears only if all technology advantage is lost. Everything depends on the right transversality condition.

3.4.4.4 Joint venture breakdown–The transversality condition

To derive the final condition we need to rethink the investor's options in the beginning. The investor can either keep the investment project himself or share his firm.

Suppose the investor employs a local partner in the beginning, $t = 0$. It follows that the investor's profit from a shared enterprise is higher than the profit from a wholly owned subsidiary. In other words, the initial cost reduction outweighs the reduction of the profit prospect due to the payment to the local firm and the diminishing technological advantage.

Naturally, this can not last. There are two mutual reinforcing powers at work which reduce the profit of the investor. First, there is the direct effect due to the profit transfer to the partner necessary to reward him for his efforts and due to the ownership arrangement.

In addition, there is an indirect effect due to learning. The smaller the technological advantage, the more important is the profit contribution from the partner. Thus, there is a rising incentive to give away by and by a larger share of the investment project. But this increases the learning effect in two ways.

The first effect is the result of the increased share. The higher the share for the local firm, the more will be learned. The second, less obvious effect comes from the decrease of the output price. A decreased price implies an increased quantity. This in turn increases the learning effect again. So one can conclude that although at $t = 0$ the profit of the joint venture is higher than for a wholly owned firm there must be some moment in time with either $A_T = 0$ or $\pi_s(T) \leq \pi(T)$. In economic terms, the technological advantage is vanished at some moment in time or the local participation is abandoned.

To formalize this intuition, consider the instantaneous profit functions $\pi_s(t)$ and $\pi(t)$. Both are increasing functions of A. Whenever a local partner is involved the technology advantage diminishes. It follows that the instantaneous profit decreases, too. This implies that whenever the profit $\pi_s(t)$ with local participation is below $\pi(t)$ it remains below it forever. It follows, that in the beginning a joint venture can arise if and only if $\pi_s(0) > \pi(0)$. Because the strict monotonicity in A of the instantaneous profit functions applies to the inter-temporal profit functions it suffices to consider the instantaneous profit functions only. Accordingly, the discounting factor can be neglected as it cancels out.

The investor is a price-taker under perfect competition. Consequently, the price path is determined fully with the determination of the technology path. This justifies the ignorance of the other features, as price and quantity, in the decision problem of the investor and, accordingly, in the profit functions, π_s and π. The technology gap, A, is affected by the share s only.

Using the instantaneous profit functions we find that the transversality condition is determined by that moment T in the time interval $[0, \infty)$ for which either

$$\pi_s(T) \leq \pi(T) \text{ if } A_T > 0 \text{ or} \tag{3.62}$$

$$A_T = 0 \tag{3.63}$$

is fulfilled.

Notice that the above intuition has the additional implication that this T is finite. Instead of taking the limit we have to seek a finite moment in time. This important insight is formulated in a proposition.

Proposition 3.5 *If initially a joint venture was formed then there exists a $T \in (0, \infty)$, so that for the final technology gap A_T two cases are possible:*

1. $\pi_s(T) = \pi(T)$ *and* $A_T > 0$ *or*

2. *there exists* $T_0 \leq T$ *so that* $A_{T_0} = 0$ *with* $\pi_s(T_0) > \pi(T_0)$.

Proof: By assumption, in $t = 0$, $\pi_s(0) > \pi(0)$. From the investor's optimization problem we know that $\dot{A} = 0$ implies $A(t) = 0$ or $s(t) = 0$. For $A(t) = 0$ there is nothing to prove here. Thus we can assume $s(t) > 0$.

We find for the technology derivative of the inter-temporal profit functions at the market equilibrium that

$$\frac{d\pi}{dA} = p_A[q_t + \frac{dq}{dp}(p_t - q_t - 1)] \geq 0 \tag{3.64}$$

$$\begin{aligned} \frac{d\pi_s}{dA} &= (1-s)p_A\left\{q_t + \frac{dq}{dp}\left[p_t - q_t - 1 - 2s_t q_t\right)\right]\right\} \\ &\geq 0. \end{aligned} \tag{3.65}$$

The profit decreases with a decreasing technology advantage.

For a joint venture to sustain, the profit of the joint venture, $\pi_s(t)$, with a technology advantage A_t at the moment t must exceed the profit from the wholly owned subsidiary, $\pi(t)$. According to (3.64) and (3.65) the profits $\pi_s(t)$ and $\pi(t)$ decrease in technology as long as $s(t) > 0$ for the joint venture and, consequently, the spillover is not zero. However, while the reduction of the profit for a wholly owned subsidiary depends on the technology change only, the profit of the joint venture depends on the local participation share, too. Because, with a decreasing technology advantage, the share given away rises[24] for a joint venture, the respective profits decline faster than for a wholly owned subsidiary with the same technology advantage. A simple comparison of the technology derivatives (3.64) and (3.65) supports the economic argument analytically.

From the continuity of the profit functions π and π_s, these must cross for some technology level $A(T) = A_T$ and some time T. As a consequence, the

[24]See proposition (3.3)

time where the profit functions cross must be finite. It is emphasized that a simple reduction in shares does not suffice to stop this process because the price reduction due to spillover has an impact on the second determinant, the quantity produced, too.

If $A_T > 0$ the final technology gap is found and the proof would be complete. But it may happen that the technology advantage has been lost before the profit of a wholly owned subsidiary can exceed that of a joint venture. This can happen because even without a technology gap the profit may be positive due to the synergistic nature of the joint venture. One ends up with a stable joint venture without technology spillover.

This can only be the case at a moment T_0 before the profit functions can cross. Otherwise, the proof would have been completed already. It follows that $T_0 \leq T$. ***q.e.d.***

Figure (3.5) illustrates the proof. It shows the time paths of the instantaneous profit functions π and π_s in the π-A-space. Because we are interested in the evolution of the profit while A decreases the coordinate system is mirrored, that is, one starts from the right and goes to the left as time passes. The time starts at the initial technology advantage, A_0, and at initial profits $\pi_s(0)$ and $\pi(0)$, respectively. It is assumed that A_0 is small enough to allow local participation, $s_0 > 0$. With A_0 the profits $\pi_s(0)$ for a joint venture exceed profits $\pi(0)$ for a wholly owned subsidiary. The time paths of $\pi_s(t)$ and $\pi(t)$ are drawn accordingly. The line A represents the time path of $\pi_s(t)$. The line D represents the corresponding time path of the profit function $\pi(t)$ for a wholly owned subsidiary.

If a joint venture was formed the technology advantage A decreases over time. So the profits of the joint ventures and the comparable wholly owned subsidiary do. Thus, a point that was placed in $(A_0, \pi_s(0))$ moves downward along the path A. Analogously, a firm that started in $(A_0, \pi(0))$ moves along the line D. Two different vertical lines denote two different possible positions of $A_T = 0$ relative to A_0.

For the vertical line B the technology exhaustion point $A_T = 0$ is located in a way that $\pi_s(T) > \pi(T)$. E_1 and E_2 denote the respective points where $\pi_s(t)$ and $\pi(t)$ cross the line B. At these point the whole technology advantage is lost. Because for $A_T = 0$ the profit $\pi_s(T)$ of the joint venture is larger than the profit of the wholly owned subsidiary $\pi(T)$ a stable joint venture persists after T. This implies for the final share $s_T > 0$. Because all technological advantage is dissipated this s is stationary afterwards. So all other quantities of interest, like prices and demand, are.

If in contrast A_0 is relatively large it can occur that $\pi_s(t) \leq \pi_t$, but $A > 0$. This is the situation where the line C describes the position of $A_T = 0$. The crossing point of the profit functions is the point F. Under these circumstances the investor buys back the whole participation share s from the local partner at the moment T yielding a persistent positive technology gap $A_T > 0$. The joint venture breaks down by a buy-out.

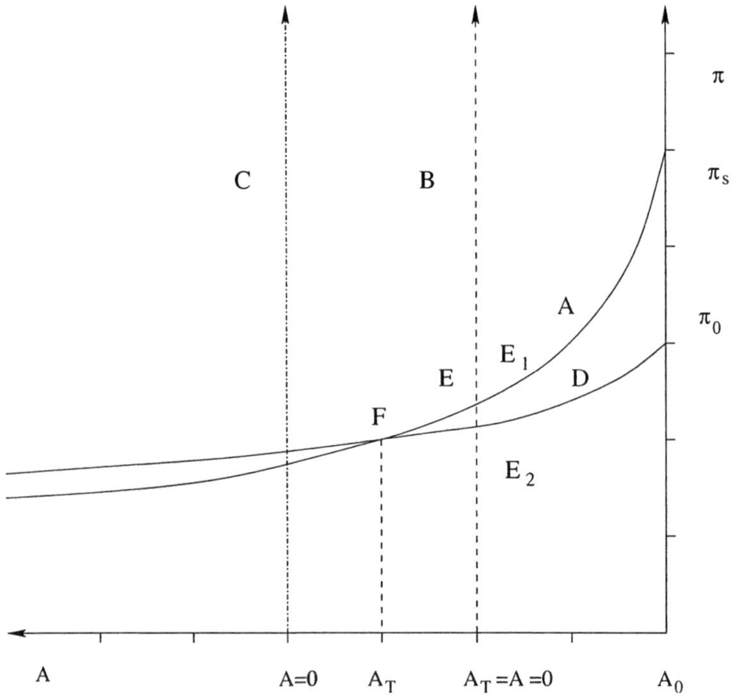

Figure 3.5: Transversality condition: The evolution of $\pi(A)$ over time

Proposition (3.5) sheds a new light on the optimal sharing behavior of the investor. In the static model it is impossible that an investor buys back the shares as the technology advantage declines. In contrary, the incentive to give away more and more shares rises. The result of the dynamic optimization as pointed out in proposition (3.5) turns this behavior upside down.

So far the final condition was described qualitatively. For an analytical description one has to find the exact transversality condition. From the proof of proposition (3.5) one can conclude that either the profit function π taken as a function in t is a final curve as long as A_t is not negative, it is $A_T = 0$ otherwise.

If one has a fixed final value, i.e. $A_T = 0$, with indeterminate final time the condition for an optimal endpoint is

$$G - \dot{A}G_{\dot{A}} = 0. \tag{3.66}$$

In case of a final curve, i.e. $\pi(T)$, the the transversality condition is given by[25]

$$G + (\dot{\pi} - \dot{A})G_{\dot{A}} = 0. \tag{3.67}$$

Again, because the transversality condition is formulated in A and $\dot{A}$ the profit functions are replaced by $G(A, \dot{A})$ if possible. The final curve π can not be properly formulated this way. Because the Euler equation is formulated without discounting, the transversality conditions has to be formulated accordingly.

We have the following proposition, that describes the feasible final states of the joint venture.

Proposition 3.6 *Suppose that A_0 is small enough that a joint venture could have been formed initially. Then at the final time T the following statements hold.*

1. *The condition (3.66) is not feasible. Thus the joint venture breaks down before the technology advantage is exhausted.*

[25] See (Klingbeil 1988), chapter 1

2. *Let* $\frac{d\pi}{dA}$ *denote the derivative of the profit of the wholly owned subsidiary with respect to the technology level. A joint venture arises and breaks down by a buy-out if for the technology derivative of the profit of the wholly owned subsidiary and the participation share* s_T *at the final time* T *holds that*

$$\left(s_T + (1 - 2s_T)\frac{d\pi}{dA}\right)\left(1 - s_T\frac{d\pi}{dA}\right) < 0. \tag{3.68}$$

3. *The investor must buy back a positive share from the local partner as long as* $\pi(t) > 0$.

Proof: First, some already known formulae are rewritten for easier use.

$$\dot{\pi} = \dot{A}\frac{d\pi}{dA}, \tag{3.69}$$

by computation, and

$$s_t = -\frac{\alpha\dot{A}}{q_t},$$

by definition of the spillover (3.42) from which we can conclude that

$$-\alpha\dot{A} = s_t q_t. \tag{3.70}$$

From the definition of G we get the expression

$$G = (1 - s_t)\pi(t) - \alpha\dot{A}(q_t + \alpha\dot{A}). \tag{3.71}$$

With the relation (3.70) this implies

$$G = (1 - s_t)\pi(t) + (1 - s_t)s_t q_t^2. \tag{3.72}$$

Furthermore, from the definitions of G and the formula (3.45) for $G_{\dot{A}}$ one derives

$$G_{\dot{A}} = \alpha\left(\frac{\pi}{q_t} - q_t - 2\alpha\dot{A}\right), \tag{3.73}$$

and in conjunction with the spillover definition in form of (3.70)

$$\dot{A}G_{\dot{A}} = -s_t\pi + (1 - 2s_t)s_t q_t^2. \tag{3.74}$$

This yields

$$G - \dot{A}G_{\dot{A}} = \pi + s_t^2 q_t^2.$$

The condition (3.66) becomes

$$\pi_0 = -s_t^2 q_t^2.$$

Because only equilibria with positive supply and non-negative profits are considered[26], this transversality condition never can be satisfied, as part 1. of the proposition claimed.

This means that no t with $s > 0$ exists for which the inter-temporal profits of the investor can be maximized and which ends up with $A_T = 0$. The only possibility left is breakdown by buy-out.

Thus, the only valid final condition can be derived from the transversality condition (3.67). We can simply add $\dot{\pi}G_{\dot{A}}$ to the transversality condition (3.66), $G - \dot{A}G_{\dot{A}}$ was already computed. From this one finds that, with $\kappa_t = 1 - 2s_t$, $\kappa \in (-1; 1]$,

$$G + (\dot{\pi} - \dot{A})G_{\dot{A}} = \left(1 - s_t \frac{d\pi}{dA}\right) \pi(t) + \left[\kappa_t \frac{d\pi}{dA} + s_t\right] s_t q_t^2 \tag{3.75}$$

delivers the correct transversality condition for this case. Setting it to zero one gets

$$\pi(t) = -\frac{s_t + \kappa_t \frac{d\pi}{dA}}{1 - s_t \frac{d\pi}{dA}} s_t q_t^2. \tag{3.76}$$

To be non-negative in the end the right-hand side of equation (3.76) requires that

$$\left(s_T + \kappa_T \frac{d\pi}{dA}\right)\left(1 - s_T \frac{d\pi}{dA}\right) < 0.$$

This is just the condition given by proposition (3.6) part 2.

However, this condition never holds for $s = 0$. The transversality condition (3.75) reveals that it can be the case for $\pi = 0$ only. Thus, as long as

[26]Remember that $q_t = 0$ is an equilibrium supply, too.

$\pi(t) > 0$ there is some share of the firm left for the local partner. Because the profit of the wholly owned subsidiary exceeds the profit of a joint venture for the final technology advantage, the investor buys back the shares.

q.e.d.

Proposition (3.6) delivers the main result for joint ventures with controllable spillover. It states that a buy-out is the only feasible exit path for a profit maximizing investor. In contrast, proposition (3.5) states that breakdown is one of two possibilities only for the joint venture to evolve. But this proposition ensured, that if a breakdown happens it happens in finite time. The possibility that technology is exhausted before the profits of the joint venture drop below the profits of a wholly owned subsidiary remained possible. But if this happens, it happens earlier than joint venture breakdown.

This last possibility was excluded by proposition (3.6). The economic intuition was already given by the conclusion that technology exhaustion happens earlier than joint venture breakdown. This implies that it is always possible to reduce the share of the local partner and increase the future profit from the joint venture. Such a procedure is possible as long as the the partial profit functions π_s and π_0 cross at a negative value. Because a rise in the profit of the joint venture rises the profit of the local partner, too, the local partner agrees.

The set of possible solutions is restricted further by part 2. of proposition (3.6). Unluckily, the condition (3.68) is not very handy because the exact solution for $\frac{d\pi}{dA}$ requires the solution of a polynomial of the fourth order in s. A positive T does not exist if the condition is not fulfilled. In other words, a joint venture would not have been formed beforehand.

As the proposition (3.6) part 3. claims, the profit path of the joint venture does not approach the profit path of the comparable wholly owned subsidiary smoothly, except for the case where $\pi = 0$. Consequently, there is a share $s_T > 0$ left in the hands of the local partner. One might wonder if a final buy-out requires a payment from the investor. However, this is not the case. The investor is always able to withdraw the joint venture totally and to set up a wholly owned subsidiary with the profit $\pi(t)$. Because this enterprise is not shared, the profit remains forever at the level for A_T the joint venture had approached at exit time. Thus an ultimatum argument is valid, where the investor pays a small amount $\varepsilon > 0$ to the local partner. At the margin, the investor pays nothing and the local partner agrees. This line of argumentation hinges on the fact that the joint venture could be set up without fix cost and no relationship-specific investments were required. The discussion of the related strategic problems if the setup would have been costly is a topic beyond the scope of this treatise[27].

Besides the economic result, proposition (3.6) yields the numerical solution for the final moment T. For the actual computation the final technology level, A_T, has to be determined simultaneously with the final moment T. The condition for A_T is that $\pi(T) = \pi_s(T)$.

The computed values for T and A_T can plugged in into the equations for the coefficients B_i, (3.60) and (3.61). The values of T an A_T can not be given in a closed form. One has to employ numerical methods instead.

3.4.5 Spillover and joint venture instability

The investigation of a dynamic model of international joint ventures demonstrated that depending on the nature of spillover stable joint ventures, joint venture buy-out, and joint venture sell-out may occur. The theory developed so far explains the whole joint venture life-cycle in a unified way and accommodates the whole range of observed behavior.

The main properties of the modeling framework can be summarized in two propositions. The first one deals with the general relationship between joint venture instability and technology spillover.

Proposition 3.7 *Technology transfer in international joint ventures occurs if and only if the joint venture is instable.*

[27]Mathematically, a positive final "scrap" value of the firm creates a "problem of Bolza" that can be easily transformed to an objective functional accessible for the indirect Euler equation approach. See (Stengel 1993), chapter 3.

Proof: Referring to the definition (3.3) of instability we remember that this means for joint ventures with two partners that

$$\dot{s} = 0. \tag{3.77}$$

Following the order in the previous proceedings, we consider the case with autonomous spillover first. Equation (3.77) is never valid if $A > 0$, because with autonomous spillover the shares of the partners are always adapted according to the comparative statics result of proposition (3.3).

Second, we consider the case for the controllable spillover. With the spillover definition (3.35) this becomes

$$\dot{s} = -\frac{\alpha}{q_t}\left(\ddot{A}q_t - \dot{A}^2\frac{dq}{dp}p_A\right) \tag{3.78}$$

from which we can conclude that this can be true only, if the differential equation

$$\ddot{A} - \dot{A}^2\frac{dq}{dp}p_A = 0 \tag{3.79}$$

is valid.

However, for controllable spillover, the solution curve for A must also pass the Euler equation (3.47). Both equations are differential equations of the second order. Equation (3.79) is nonlinear, and equation (3.47) is linear. Due to the uniqueness of the solution of a differential equation[28] the solving functions of the two equations can not coincide. Neither for autonomous spillover nor for controllable spillover the equilibrium share s is stationary. ***q.e.d.***

The second, more specific result summarizes the relation between joint venture buy-out and the controllability of spillover in joint ventures.

Proposition 3.8 *An international joint venture ends up by a buy-out if and only if the spillover is controllable.*

[28]See (Stepanov 1956), chapter 2 No. 2

Proof: Proposition (3.6) part 1. and part 2. establish the result that a controllable joint venture must end up by a buy-out. The reverse follows from the observation that only with a decrease of the share of the local partner a buy-out is possible. For autonomous spillover the share of the local partner rises. Hence, if the share of the local partner decreases over time the spillover is controllable. ***q.e.d.***

3.5 Discussion

The theory of technology transfer in international joint ventures has several implications. We start with the presentation of testable predictions derived from the model and give explanations for previously observed phenomena, respectively.

Subsequently, a measurement method of the value of transferred knowledge is presented and applied to selected countries from eastern and central Europe. Policy recommendations and a conclusions finish the discussion.

3.5.1 Ownership and spillover

Two different types of joint ventures can be combined with respect to a prediction of the relationship between ownership shares and technology advantage. Both, the stable joint venture and the joint venture with autonomous spillover, are subject to the following prediction.

Hypothesis 3.1 *Suppose, that a joint venture is stable, or that the spillover in an instable joint venture is autonomous. Then the smaller the technology advantage of the investor the larger is the ownership share of the local firm at a fixed date.*

This hypothesis is the result of proposition (3.3) for stable joint ventures and of proposition (3.4) for instable joint ventures with autonomous spillover. The prediction applies for a cross-section of joint ventures as the temporal qualification of hypothesis (3.1) indicates.

If a spillover is present it is possible to make predictions concerning the correlation of changes of ownership shares of international joint ventures and changes of the technology advantage of the investor.

For autonomous spillover the following prediction applies:

Hypothesis 3.2 *Suppose the spillover is autonomous, that is, the ownership share of the investor of the joint venture has decreased. Then the technology advantage of the investor has decreased as well.*

For joint ventures with controllable spillover the prediction is diametrical.

Hypothesis 3.3 *Suppose the spillover is controllable, that is, the ownership share of the investor of the joint venture has increased. Then the technology advantage of the investor has decreased, too. This effect is expected to be small.*

The difference between the hypothesis (3.2) and (3.3) is not that the technology advantage has decreased. Technology dissipates in both situations. However, if the spillover is autonomous the correlation between the change of the ownership share of the investor and the change of the technology gap should have a positive sign. For controllable spillover this sign should be negative, if the effect is observable at all.

The hypothesis (3.2) is again the result of proposition (3.4). The hypothesis (3.3) follows from proposition (3.6).

In contrast to hypothesis (3.1), the hypothesises (3.2) and (3.3) consider the relationship of changes of the ownership shares and of the technology gap. They deliver inter-temporal predictions. Consequently, they can be tested with longitudinal or panel data only. Accordingly, if a panel is pooled hypothesis (3.1) is applicable only.

3.5.2 Aggregate ownership patterns

At an aggregate level information of changes of the ownership of individual firms can not be extracted in general. Thus, one can not expect predictions similar to those of the previous section. At an aggregate level the share of investment projects that are minority owned, majority owned, and wholly owned remains quite stable over time for investments in industrialized countries compared to the dynamics in central and eastern Europe, as a comparison of tables (3.4) and (3.5) suggests. Figures (3.6) and (3.7) display the data graphically. Nevertheless, even for the relatively stable situation of foreign direct investment in the industrialized countries the dynamics might be considerable at the firm level. Firms enter and exit the pool of firms that are observed as an aggregate. They may change the ownership distribution among the joint venture participants. Without individual information conclusive predictions are impossible.

However, for the former socialist countries in central and eastern Europe the situation is different. Because the initial date for the entry of firms was observed, one can, based on the model of the joint venture life-cycle, make predictions about the evolution of the pattern of ownership at the aggregate level.

The aggregate pattern of ownership is influenced by three factors. Firms enter the pool of observed firms. Firms leave the pool. Firms that remained in the pool change their participation share.

However, after the initial moment the number of firms that leave the host country immediately after investment is negligible. Hence, one can ignore that firms leave the pool of observation. Only those firms that enter the foreign country and firms that were already observed influence the aggregate ownership pattern.

Thus at the aggregate level one can state

Hypothesis 3.4 *If the spillover is controllable for most firms that enter the country, the share of firms that are wholly owned subsidiaries and majority owned joint ventures rises over time initially.*

This hypothesis is a conclusion from the result that a joint venture with controllable spillover breaks down by buy-out, as proposition (3.6) suggests. According to the empirical results presented in table (3.1) we assume that for most joint ventures the spillover is controllable. The share of firms who are wholly owned is expected to rise over time. For firms that increased the share of the investor, but are not wholly owned, the adaption process towards a wholly owned subsidiary is not finished yet. Nevertheless, the share of majority owned joint ventures should rise at the aggregate level accordingly.

Under the controllable spillover assumption entry is the only source for minority owned joint ventures. In contrast, wholly owned subsidiaries and majority owned joint ventures have the internal dynamics within the pool of observed firms as an additional source for an increase of their quantity. Consequently, the share of wholly owned subsidiaries and majority owned joint ventures among all foreign direct investment projects should rise.

Of course, it is assumed that firms that enter do not change their initial behavior. This implicitly yields that the relative technology differential remains constant. Because the spillover effect is small if the spillover is controllable, this implicit assumption does not pose a problem.

For Germany-based investment projects in the transition economies table (3.5) shows the aggregate pattern of ownership over time. The share of joint ventures is considerably higher in the period from 1990 to 1994. It approaches an approximately stable level afterwards.

A similar observation can be made for minority and majority owned joint ventures. The share of firms that are majority owned joint ventures rises initially but becomes stable afterwards. The stabilization of the shares of the different ownership types after an initial adaption period indicates that the exit of firms becomes influential for the aggregate ownership pattern.

In contrast, the share of minority owned joint ventures decreases before it stabilizes.

We can observe in table (3.5) that initially the majority of firms own 50 % and less of their investment project. In 1998 less than 20 % of all firms that have a foreign direct investment hold a minority share or half of the investment project. The initial pattern is reversed. If exits are

Year	min	%min	half	%half	maj	%maj	wos	%wos
1990	694	15.60	170	3.82	1175	26.40	2411	54.18
1991	687	14.56	193	4.09	1194	25.30	2646	56.06
1992	697	14.29	198	4.06	1215	24.90	2769	56.75
1993	624	13.16	203	4.28	1131	23.85	2784	58.71
1994	623	12.81	187	3.85	1111	22.85	2942	60.50
1995	683	12.90	183	3.46	1191	22.50	3236	61.14
1996	657	12.00	208	3.80	1156	21.11	3454	63.09
1997	656	11.56	208	3.66	1289	22.71	3524	62.08
1998	616	10.50	180	3.07	1273	21.69	3800	64.75

Table 3.4: Time pattern of aggregate ownership distribution for Germany-based foreign direct investments in manufacturing in industrialized countries (Absolute numbers and percentages of firms of all FDI projects registered for the particular year, min=minority joint venture, investor holds $20 \leq, < 50\%$, half=investor holds 50 % percent of the joint venture, maj=majority joint venture, investor holds $50 <, < 100\%$ of the joint venture, wos=wholly owned subsidiary, investor owns the whole investment project); Data Source: Deutsche Bundesbank

negligible one can conclude that the pattern of ownership changes according to hypothesis (3.4).

It may be argued that this observation is the result of an adaption process of the investor to the conditions in the foreign country. The idea is that investor starts with minority joint venture. But later on, as the investor has become more experienced, the investor prefers to enter the same country with a majority owned joint venture or with a wholly owned subsidiary.

This argument is partly a pro and partly a con for the model developed herein. It is a pro because the model of the joint venture life cycle is intended to explain the adaption process in international investment projects driven by learning. It is a con because it is not the investor that learns about the host country in this model. It is the local firm that learns the foreign technology.

This allows a decisive hypothesis to test this argument. It is testable

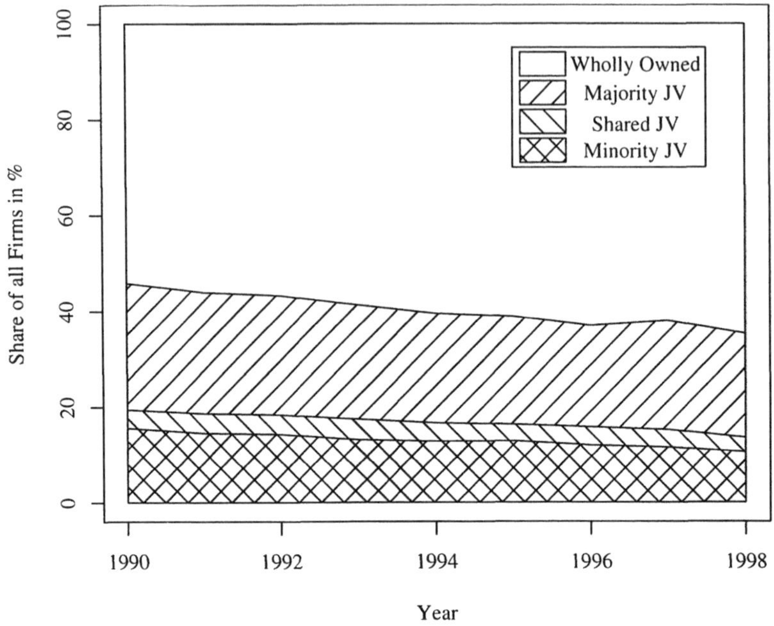

Figure 3.6: Time pattern of aggregate ownership distribution for Germany-based foreign direct investment in manufacturing in industrialized countries

with firm-level data only.

Hypothesis 3.5 *Suppose that an investor enters the same foreign country repeatedly. Then the investor learns from the host country if the ownership share of the investor of every follow-up investment rises. If the ownership share of the local partner rises for the same setting the local partner learns the foreign technology.*

This hypothesis allows an assessment if the host country learns from the investor, or if the investor learns from the host country. Hypothesis (3.5) provides a policy monitoring tool for the assessment of the effectiveness of technology transfer by foreign direct investment, too.

Year	min	%min	half	%half	maj	%maj	wos	%wos
1990	61	53.04	13	11.30	34	29.57	7	6.09
1991	88	42.11	21	10.05	82	39.23	18	8.61
1992	96	27.75	24	6.94	152	43.93	74	21.39
1993	95	19.00	35	7.00	226	45.20	14	28.80
1994	129	17.82	37	5.11	309	42.68	249	34.39
1995	189	17.68	42	3.93	410	38.35	428	40.04
1996	235	16.86	64	4.59	533	38.24	562	40.32
1997	264	16.26	66	4.06	625	38.49	669	41.19
1998	275	14.38	78	4.08	715	37.38	845	44.17

Table 3.5: Time pattern of aggregate ownership distribution for Germany-based foreign direct investments in manufacturing in transition countries; Legend as in table (3.4); Data Source: Deutsche Bundesbank

3.5.3 Technology spillover in foreign direct investment

The model developed herein is able to explain the mixed and confusing evidence that was observed in answering the question if foreign direct investment is a technology transmission device.

From the tables (3.4) and (3.5) in the previous section one could discover that the share of firms that possess wholly owned subsidiary rises over time. Thus, the share of firms that transfer technology declines.

Consequently, one can sometimes observe technology transfer by FDI and sometimes not. It depends on the relative technology levels of the firms entering the market and the share of firms that generate spillover yet. For example, (Konings 2001) considers the technology differential between international joint ventures and local firms in the host countries Bulgaria, Romania and Poland. Except for Poland, a significant differential could not be found. In such a case a spillover can not be expected. There is nothing to learn.

On the other hand, (Smarzynska 2002) finds that Poland is apparently preferred by firms with a large technology advantage. But, as the model tells us, in this case a spillover does not occur. Because the firms have

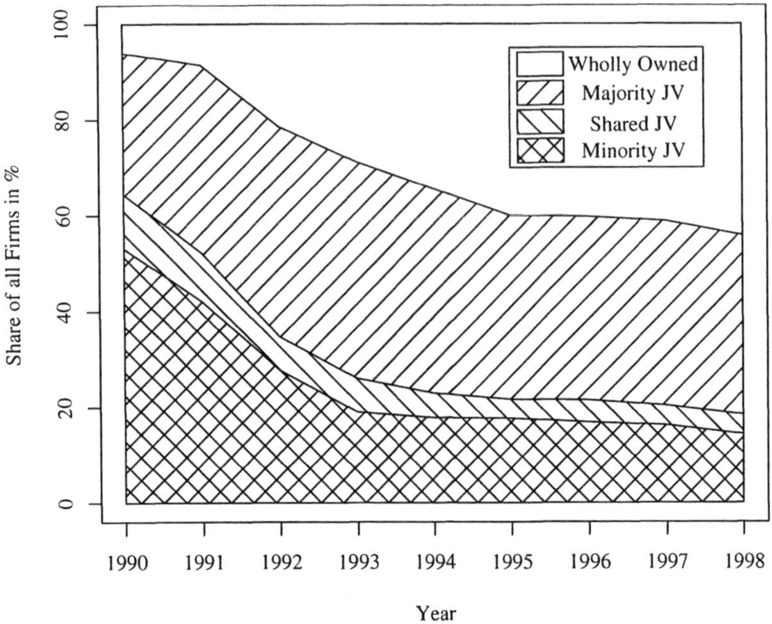

Figure 3.7: Time pattern of aggregate ownership distribution for Germany-based foreign direct investments in manufacturing in transition countries

a large technology advantage they prefer to enter the host country as a wholly owned subsidiary. (Smarzynska 2002) confirms even this.

For intermediate technology gaps a small spillover occurs but the joint venture breaks down by a buy-out. Afterwards technology is not transferred any more. This leaves firms freshly entering the market as candidates for technology transfer only.

Because both, very high and very small technology differentials, may undermine the occurrence of technology spillover, an unambiguous statement can not be given. A careful inquiry is necessary to find a verdict which branch and which country may enjoy spillover or not. So one might flee into vagueness with the saying, 'It depends'.

More valuable than that a measure of the value of transferred knowledge would be. It can be easily constructed with the model at hand, as the following section will show.

3.5.4 The value of transferred knowledge

In the discussion of the mixed results with respect to the spillover from FDI the notion of absorptive capacity was introduced[29]. Although the idea that not all available knowledge can be used is an appealing idea, the concept in itself remained quite vague.

In the context of the model, it can be given an empirically feasible formulation. Remember, that the initial technology advantage was denoted A_0 and the final technology advantage was denoted A_T.

Then the absorptive capacity of an industry in the host country is, according to the spillover specification (3.35),

$$\int_0^T \dot{A} dt = -\beta \int_0^T s_t q_t dt. \tag{3.80}$$

This can be equivalently written as

$$A_T - A_0 = -\beta \int_0^T s_t q_t dt. \tag{3.81}$$

Note, that both D and s are functions of time by virtue of the technology path $A = A(t)$. According to these formulae, one can express the extent of absorbed technology as a change in sold quantities and participation shares.

Now one can approximate the integral[30] on the right-hand side of equation (3.81) by the following empirically meaningful equation

$$A_T - A_0 = -\beta T \frac{s_T q_T + s_0 q_0}{2}. \tag{3.82}$$

For controllable spillover, where the joint venture was abandoned this becomes

$$A_T - A_0 = -\beta T \frac{s_0 q_0}{2}. \tag{3.83}$$

[29](Kokko 1994) was to the author's knowledge the first article to introduce this idea. The other pioneering work is (Keller 1996). (Keller 2001) reviews this facet of technology transfer, too, and gives some references.

[30]The integral mean-value theorem (Bronstein and Semendjajev 1979), 3.1.7.2 No. 8, with a linear approximation of the time path of $s_t q_t$ was used.

Because raw quantities are usually not comparable, it is useful to replace the quantities in formulae (3.82) and (3.83) with the respective sales figures. This delivers the value of technology that was absorbed. However, with sales figures one has to take care of price effects if long periods are considered.

The only unobservable variable of the right-hand side is the parameter β. It is a positive number and may be estimated or even guessed according to plausibility.

For retrospective data formulae (3.82) and (3.83) deliver a method to measure the value of technology that was absorbed.

Some approximate absorptive capacities valued with market prices are computed from the dataset (Universität München 2002)[31]. Table (3.6) provides the results for the value of absorbed knowledge, i.e. realized absorption capacities, for selected eastern European countries.

Country	Overall Number of Projects	N	Average breakdown time in years	Absorbed knowledge value per firm in Mio. DM	Absorbed knowledge value per firm per year in Mio. DM
Poland	272	20	3.57	8468.61	2371.21
Russia	117	6	5.50	1274.29	231.69
Czech Rep.	257	25	5.04	37487.38	7437.97
Hungary	168	17	6.44	1369.56	212.52

Table 3.6: Average joint venture breakdown times and the value of absorbed knowledge for selected eastern European countries; Data Source: (Universität München 2000), own computations

The parameter β was assumed to be 1. It is plausible because it can be expected that the determinants of learning in central and eastern Europe are not too different from Germany. Moreover, the sales and prices were assumed to remain constant. This assumption is a tribute to the lack of observations of sales figures in the beginning of the investment project.

[31]It is a slightly extended version of the dataset used in chapter 2.

Because we are interested in the absorptive capacities to explore the relationship between joint venture instability and technology transfer, only investment projects were considered that started as a joint venture but broke down by buy-out. Accordingly, formula (3.83) has to be employed.

The countries were selected according to the number of investment projects they accommodate. Only the countries with the largest number of investment projects were selected.

Altogether, 1176 investment projects were reviewed for the dataset. The numbers of overall investment projects that are hosted in the respective countries are given in the second column of table (3.6). From these 814 cases only those firms were selected that entered as a joint venture but broke down by buy-out. The number of these firms is given in column N.

The average breakdown time is of interest, because it gives an impression of the adaption speed for the respective country. In addition, according to formula (3.83), the time to breakdown is needed for the computation. That's why, the average time to breakdown gives hints for the absorptive abilities of the country, too.

The actual numbers for the value of absorbed knowledge are contained in the last two columns of table (3.6). One gives the value of knowledge absorbed per firm. The other is the value per firm per year.

The relatively low value of the absorbed knowledge per firm for the country with the highest average time to breakdown, Hungary, should not come as a surprise after the discussion of the model at hand. As a rule of thumb, the model says that the more instable a joint venture the more knowledge is transferred. The case of Hungary supports this view.

3.5.5 Policy recommendations

In general, two different types of policy applications exist. One type of policy applications tries to give hints how an optimal policy derived from theory should look like. The less obvious, other type tries to investigate the applicability of theoretical results and monitors the outcome of policy measures. Usually, the first type of applications is more spectacular, the second one is more useful.

The model of technology transfer by joint ventures developed herein can contribute to both aspects of policy recommendations with respect to foreign direct investment in general and with respect to joint ventures in particular.

For the assessment of a policy with respect to technology transfer by foreign direct investment and international joint ventures the measurement of absorptive capacities plays an important role. A government that plans to liberalize capital controls with the purpose of technology transfer can employ the methods to measure the value of absorbed knowledge to evaluate the results. Regular monitoring of the technology transfer status allows the government to correct for possibly unintended investments.

With respect to proposing purpose of policy recommendations the results derived by the model imply seemingly paradoxical hints. At least, they are paradoxical because they contradict traditional views at international investments and turn them upside down.

Suppose, one is in the position of a minister for economic affairs in a developing country and would like to promote technology transfer.

The best measure is to let the joint venture participation adapt freely. Otherwise, if a local participation share is mandatory, the firm for which a smaller local participation share is advantageous does not invest at all. Furthermore, if spillover occurs but the investor must adapt his participation share in order to fulfill a mandatory local participation share requirement the investor would withdraw his investment before the maximal possible extent of technology is transferred.

In addition, if only stable joint ventures are considered advantageous this implies that joint ventures are supported that do not transfer technology. This is a simple corollary of proposition (3.7). To formulate it provocative: to promote technology transfer it is necessary to support instable joint ventures.

On the other side, for a country that wants to keep a technology advantage the recommendations bear similarly paradoxical elements. The most counter-intuitive recommendation is to promote international investment for firms with a very large technology advantage. If the host country does not have legal participation requirements the investor will enter the foreign country with a wholly owned subsidiary only.

The second hint is intuitive. Foreign direct investment should be allowed for firms with a very low general technology level, too. But, it should be avoided that firms with an intermediate technology level make foreign direct investment in less developed countries. In this case, it can be expected that the investing firms form a joint venture with a local partner but are not able to control the technology flow. As an example, one could think of electronic consumer goods "made in China".

Finally, if firms enter a foreign market but the technology is difficult to imitate there is no need to supervise these firms. Although a technology flow may happen, the controllability of the spillover prevents a total loss of the technology advantage. Car manufacturing is a plausible example.

In general, the mixed empirical results with respect to technology transfer by foreign direct investment in connection with the model of the joint venture life-cycle suggest that international investment is inferior for the purpose of technology transfer compared to the flow of actual goods and services. The crucial issue for this verdict is the ability of firm owners to control the flow of their intellectual property.

3.6 Conclusion

This chapter dealt with the problem of technology transfer via foreign direct investment and the role that joint ventures play. Contrary to previous modeling approaches a model in continuous time is introduced. Independence of joint venture formation of the market structure is shown.

The model allows to explain the whole range of observable ownership dynamics in foreign direct investment projects. Within a unified framework buy-outs, as well static participation, or a s rise in the participation of a local partner is explained.

The development of a model of technology transfer in joint ventures based on a microeconomic disequilibrium model in continuous time overcomes several shortcomings of previous models of the joint venture lifecycle. It may come as a surprise that neither information asymmetries nor a particular market structure are necessary to explain the rise and breakdown of joint ventures as the previous literature on this topic suggested.

Of course, a newly introduced model is required to be tested, before it can be accepted. To this end, testable predictions are extracted from the model.

The approach developed herein explains several empirical observations (or their failure). These are the aggregate pattern of ownership in foreign direct investment projects and the mixed evidence of a technology transfer through foreign direct investment.

A measurement method for the extent of technology transfer is derived and applied to foreign direct investment of German firms in central and eastern Europe.

The derived policy measures are quite paradoxical. If one wants to promote aggregate technology spillover by foreign direct investment one should support instable joint ventures and avoid the stable ones. If one wants to avoid technology diffusion by foreign direct investment firms with a very large technology advantage are as appropriate as firms without any technology advantage.

Unluckily, the central message with respect to the technology transfer via foreign direct investment is a negative one. It can not be expected that technology is easily transferred if the spillover is to some extent under the discretion of the investor. On the other hand, it is shown that for autonomous spillover technology can be fully absorbed. However, as it can be concluded from the data, joint ventures with autonomous spillover seem to be an exception.

However, the measurement method proposed herein provides a simple procedure to evaluate empirically the success of technology transfer.

Chapter 4

Concluding remarks

The fundamental conclusion that can be drawn from the topics discussed in this thesis can be easily subsumed under the old saying from economic folklore: "There is no free lunch".

For the test of the sovereign risk theory in chapter 2, one finds some degree of freedom for the government to levy taxes on firms owned by a foreign investor. While this may promote local participation, as it is intended, it reduces the incentive of the investor to transfer technology. How this trade-off is solved is a subject of the social discourse. It is a topic beyond the scope of economic analysis.

The idea that foreign direct investment is a device of technology transfer and should be promoted accordingly becomes questionable in the light of the continuous-time model of the joint venture life-cycle developed in chapter 3. The trade-off that a society faces in this model is whether technology transfer or the presence of foreign capital is preferred. Technology transfer comes at the price of a loss in participation of local firms in the technologically advanced foreign investment project. The ability of firm owners to restrict the flow of intellectual property drives this outcome.

Altogether, the results suggest that economic reasoning can only provide insights into the balances one has to account for. Unambiguous advice is not possible.

The loss in naivete with respect to the interrelations between firm ownership, technology transfer and governmental power is balanced against new tools and insights that could be gained by exploration of these topics.

The development of the conditional Tobit approach for the test of the sovereign risk theory allows a more instructive and detailed insight into the determinants of ownership of foreign direct investment projects than standard methods provide. Moreover, applying the conditional Tobit approach to other questions with similar structure might add a new tool to the micro econometric toolkit.

Similarly, the reformulation of the relations between ownership and knowledge transfer in the model of the joint venture life-cycle provides a new approach to the measurement of knowledge flows. Moreover, it emphasized the role of ownership as a vehicle of control of knowledge flows.

As in the respective chapters was pointed out, the ideas developed herein have opened new questions in answering the old ones. But possibly this is the best result that can be achieved. Or, as the America economist Thorstein Veblen pointed out: "So it is something of a homiletical commonplace to say that the outcome of any serious research can only be to make two questions grow where one question grew before."[1]

[1] See (Veblen 1908)

Appendix A

Maximum likelihood theory and numerical optimization

A.1 Maximum likelihood theory

The following theorems are given for easy reference. They describe standard properties of maximum likelihood estimators.

Theorem A.1 *Let $\hat{\theta}$ be a maximum likelihood estimator for the unknown parameter vector θ in the probability density function $f(x|\theta)$ and the observations x. If f has certain regularity properties*[1] *then a series of maximum likelihood estimators $\{\hat{\theta}_T\}_{T=1,2,\ldots}$ with the corresponding likelihood function $\mathcal{L}_T$ converges in distribution to a normal distribution with the expectation θ and the variance equal to the invese of the Fisher information matrix.*

$$\sqrt{T}(\hat{\theta}_T - \theta) \longrightarrow N\left\{0, -\lim_{T\to\infty} E\left[\frac{1}{T}\frac{\partial^2 log\mathcal{L}_T}{\partial\theta\partial\theta'}\right]^{-1}\right\} \tag{A.1}$$

Proof: see (Amemiya 1985) Theorem 4.2.4 and 4.1.3 ***q.e.d.***

[1]see (Amemiya 1985), Theorem 4.1.3

Theorem A.2 *Under the regularity conditions from theorem (A.1) for the corresponding likelihood function $\mathcal{L}$ the following identity holds*

$$E\frac{\partial^2}{\partial\theta\partial\theta'}\mathcal{L} = -E\frac{\partial log\mathcal{L}}{\partial\theta}\frac{\partial log\mathcal{L}}{\partial\theta'} \tag{A.2}$$

Proof: see (Mood, Graybill, and Boes 1974), p.320 ***q.e.d.***
This result is particular important with respect to numerical applications as we will see later. It states that the Hessian of the log-likelihood function can be replaced by the outer product of the gradients of the log-likelihood function. This allows to circumvent the analytical derivation of the Hessian of the log-likelihood function. This may be advantageous if it is sufficient complicated.

A.2 Numerical optimization

Because the estimation approach is highly nonlinear the solution of the maximum likelihood requirement can not be written in analytical form. Thus one has to employ numerical procedures.

The (unrestricted) optimization problem used herein will be formulated as

$$\min_{\theta} S(\theta, X). \tag{A.3}$$

$S(\theta, x)$ is the objective function that depends on the parameters $\theta \in \mathbb{R}^m$ of interest and the observation matrix $X \in \mathbb{R}^{m \times k}$.

The general numerical optimization procedure is a stepwise approach of the general form

$$\theta_{n+1} = \theta_n + \gamma P\frac{\partial}{\partial\theta}S(\theta_n, X), \tag{A.4}$$

where $\theta_i, x \in \mathbb{R}^m$ $i = 0(1)N$ denotes the estimator after the i-th step, P is a square matrix of full rank and γ denotes the step width.

Definition A.1 *The algorithm A.4 is denoted gradient method if $P = \mathcal{E}$, where $\mathcal{E}$ is the unit matrix.*

Definition A.2 *The algorithm A.4 is denoted Gauss-Newton-method if*

$$P = -\left[\frac{\partial^2}{\partial\theta\partial\theta'}S\right]^{-1}. \tag{A.5}$$

Definition A.3 *The algorithm A.4 is denoted the method of scoring if*

$$P = -\left[E\frac{\partial^2}{\partial\theta\partial\theta'}S\right]^{-1} = I(\theta)^{-1}. \tag{A.6}$$

Prima facies it seems surprising that the method of scoring should be a more convenient base for an estimation method than the Gauss-Newton method. Why should the use of the expectation of the Hessian of the log-likelihood function produce an easier approach than the use of the Hessian itself ?

The answer give the already stated Theorem (A.2). It states that we can replace the expectation of the Hessian with the expectation of the outer product of the gradient of the likelihood function. Given that the second derivative is sufficient complicated, as it is in our case, this is indeed an easier approach.

This idea was taken up in (Berndt, Hall, Hall, and Hausman 1974). Therein was proven that the approach with an estimated Fisher-Information is asymptotically equivalent to the method of scoring. Their procedure was used in the implementation.

Definition A.4 *The algorithm A.4 is denoted BHHH if*

$$P = -\left[E\frac{\partial S}{\partial\theta}\frac{\partial S}{\partial\theta'}\right]^{-1}. \tag{A.7}$$

A.3 Implementation issues

So far we have collected the necessary tools for estimation. Now we can start some implementation details.

The most important implementation issue is the exploitation of some structural model properties to get a reasonable fast computation.

The implementation was done in native **R** [2]. This makes the computation quite slowly per se. As (Venables and Ripley 2000) report the speed of native **R** is about 15-20 times slower than the corresponding **C** implementation.

On the other hand the problem at hand is rather involved so that the lucidity of the high level programming is considered advantageous compared to the relative low level approach via **C**.

Nevertheless there are some possibilities to improve the computation performance even in **R**. The source for most of the optimizations come from the vectorization of the computation problem. That is the replacement of loop with the corresponding vector operations.

Again, as reported in (Ihaka and Gentleman 1996) vectorization increases the speed of native **R** about 3- 5 times. So the design problem comes down to the question how the formulae above can be organized that the advantages of vectorization can be exploited.

A.4 Implementation

```
###################### Helpers #########################

##################### zetify ###########################
# standardization of observations
#x matrix of observations
#beta vector of coefficients
#

zetify<-function(u,x,beta,sigma){

return((u-x %*% beta)/sigma)
}

############ hazardify ##################################
# returns hazard rate  of standardized observation
# with respect to standard normal distribution
```

[2]see (Ihaka and Gentleman 1996) on the **R** statistical computing environment

```
hazardify<-function(zet){
  return(dnorm(zet)/(1-pnorm(zet)))
}

##################### End Helpers #######################

##################### Derivatives and Scores ############
# y response vector
# x vector of influences
# beta estimated coefficients
# sigma estimated variance
# cl lower bound
# cu upper bound

zetmatrix<-function(y,x,beta,sigma,cl,cu){
 a.zetmatrix<-function(y,x,beta,sigma,cl,cu){
  x.nrow<-dim(x)[1]
  x.ncol<-dim(x)[2]
  a<-matrix(nrow=x.nrow,ncol=x.ncol+1)
  for(i in seq(1,x.nrow)){
   for(j in seq(1,x.ncol)){
    a[i,j]<--hazardify(zetify(cl,x[i,],beta,sigma))*x[i,j]/sigma
   }
   a[i,x.ncol+1]<--hazardify(zetify(cl,x[i,],beta,sigma))
                  *zetify(cl,x[i,],beta,sigma)/(2*sigma**2)
  }
 a
 }

 b.zetmatrix<-function(y,x,beta,sigma,cl,cu){
  x.nrow<-dim(x)[1]
  x.ncol<-dim(x)[2]
  b<-matrix(nrow=x.nrow,ncol=x.ncol+1)
  for(i in seq(1,x.nrow)){
   for(j in seq(1,x.ncol)){
    if(y[i]<cu){
     b[i,j]<-zetify(y[i],x[i,],beta,sigma)*x[i,j]/sigma
    }
    else{
     b[i,j]<-hazardify(zetify(cu,x[i,],beta,sigma))*x[i,j]/sigma
    }
   }
   if(y[i]<cu){
    temp1<-(zetify(y[i],x[i,],beta,sigma))
    b[i,x.ncol+1]<--1/(sigma**2)+(temp1*temp1)/(2*sigma**2)
   }
   else{
```

```
    b[i,x.ncol+1]<-hazardify(zetify(cu,x[i,],beta,sigma))
                  *zetify(cu,x[i,],beta,sigma)/(2*sigma**2)
   }
  }
  b
 }
 Z<-a.zetmatrix(y,x,beta,sigma,cl,cu)+
    b.zetmatrix(y,x,beta,sigma,cl,cu)
 grad<-function(y,x,beta,sigma,cl,cu){
   g<-vector("numeric",length(beta)+1)
   for(i in seq(1,length(beta)+1)){
   g[i]<-sum(Z[,i])
  }
  g
 }
 #
 # first argument: z-matrix
 #  second argument: gradient of log-likelihood
 #
 gradient<-grad(y,x,beta,sigma,cl,cu)
 list(Z,gradient)
}

################### log likelihood for a single observation
#
# y number, x p-vector for which log likelihood should be maximized
# beta estimated coefficients
# sigma estimated variance
#

fun.logmaxlik<-function(beta,sigma,y,x,cl=0,cu=1){
 fun.unilogmaxlik<-function(beta,sigma,y,x,cl,cu){
  A<- -log(1-pnorm(zetify(cl,x,beta,sigma)))
  if(y<cu){
   B<- log(dnorm(zetify(y,x,beta,sigma))/sigma)
  }
  else{
   B<-  log(1-pnorm(zetify(cu,x,beta,sigma)))
  }
  return(A+B)
 }
 logmax<-vector("numeric",0)
 for (j in seq(1,dim(x)[1])){
  logmax<-c(logmax,fun.unilogmaxlik(beta,sigma,y[j],x[j,],cl,cu))
 }
 return(logmax)
}
```

```
################ fun.full
# full log likelihood
#
fun.full<-function(beta,sigma,y,x,cl=0,cu=1){
 sum(fun.logmaxlik(beta,sigma,y,x,cl,cu))
}

################ r2mf
# mc fadden measure
#
r2mf<-function(y,x,beta,sigma,cl=0,cu=1){
 ymean<-mean(y)
 yvar<-var(y)
 print(ymean)
 print(yvar)
 fun.ymeanlogmaxlik<-function(y,ymean,yvar,cl,cu){
  A<- -log(1-pnorm((y-ymean)/yvar))
  if(y<cu){
   B<- log(dnorm((y-ymean)/yvar)/sigma)
  }
  else{
   B<-  log(1-pnorm((y-ymean)/yvar))
  }
  return(A+B)
 }
 logmax<-vector("numeric",0)
 for (j in seq(1,dim(x)[1])){
  logmax<-c(logmax,fun.ymeanlogmaxlik(y[i],ymean,yvar,cl,cu))
 }
 lnull<-sum(logmax)
 lhat<-fun.full(beta,sigma,y,x,cl,cu)
 cat("lnull\n")
 print(lnull)
 cat("lhat\n")
 print(lhat)
 return(1-lhat/lnull)
}

################## r2mz ##################################
# McKelvey Zavoina measure
#

predict<-function(x,beta,cl,cu){

 predict.one.unrestricted<-function(x,beta,cl,cu){
  return(x%*%beta)
 }
```

```
 predict.one.restricted<-function(x,beta,cl,cu){
   if(x%*%beta>cu) {
    return(cu)
   }
   return (x%*%beta)
 }
 yhat<-vector("numeric",dim(x)[1])
 predict.unrestricted<-function(x,beta,cl,cu){
  for(j in seq(1,dim(x)[1])){
   yhat[j]<-predict.one.unrestricted(x[j,],beta,cl,cu)
  }
  return(yhat)
 }
 predict.unrestricted(x,beta,cl,cu)
}

r2mz<-function(y,x,beta,sigma,cl,cu){
 ypred<-predict(x,beta,cl,cu)
 ydev<-ypred-mean(ypred)
 n<-length(y)
 return((t(ydev) %*% ydev)/(t(ydev) %*% ydev +n *sigma) )
}

##################### Estimation #########################
#
# start estimator
#
###########################################################
#
# start estimator: only values below c upper are used
#
cs.start<-function(y,x,cu=1,cl=0){

 bet0<-vector("numeric",dim(x)[2])
 ystar<-y[y<cu]
 xstar<-x[y<cu,]
 if ( length(ystar) / length(y) < 0.02){
  stop("Sorry, not enough Variation in y")
 }
 bet0<-lm(ystar~xstar-1)$coeff
 var0<-summary(lm(ystar~xstar-1))$sigma
 return(list(bet0,var0))
}
#
# scorestep
#
# returns 1. next estimation
#         2. value of log likelihood
```

```
#         3. change of log likelihood (for break conditions)
# ist die schrittweite
scorestep<-function(y,x,beta,sigma,cl=0,cu=1,step=1/2,rep=5,brk=1000){
#
# starters
#
 blen<-length(beta)
 fvaluealt<-0
 fvalue<-fun.full(beta,sigma,y,x,cl,cu)
 betaalt<-beta

 betaalt<-c(betaalt,sigma**2)
 i<-0
 fvalkeeper<-vector("numeric",1)
 fvkeeper<-fvalue
 while ((abs(fvalue-fvaluealt)>1/brk)&&(i<rep)){
  i<-i+1
  zet<-zetmatrix(y,x,betaalt[1:blen],sqrt(betaalt[blen+1]),cl,cu)
  score<- (t(zet[[1]]) %*% zet[[1]])
  betanext<-betaalt+step*(solve(score) %*% zet[[2]])
  fvaluealt<-fvalue
  fvalue<-fun.full(betanext[1:blen],sqrt(betanext[blen+1]),y,x,cl,cu)
  fvkeeper<-c(fvkeeper,fvalue)
  betaalt<-betanext
 }
#
# building the  return frame
#
 num.iterations<-i
 laststep<-abs(fvalue-fvaluealt)
 lastmaxlik<-fvalue
 coeff<-betanext[1:blen]
 rsvar<-betanext[blen+1]
 rsvartilde<-rsvar*(length(y)-1)/(length(y)-blen)
 stderr<-sqrt(rsvar)
 infomat<-solve(score)
 infomattilde<-rsvartilde*solve(score)
 tvalues<-coeff/sqrt(as.vector(diag(infomat[1:blen,1:blen])))
 list(
  coeff=coeff,
  tvalues=tvalues,
  sigma2=rsvar,
  r2mz=r2mz(y,x,coeff,rsvar,cl,cu),
  stderr=stderr,
  sigmatilde=rsvartilde,
  Information=infomat,
  Informationtilde=infomattilde,
  logmaxlik=lastmaxlik,
```

```
  lastimp=laststep,
  iterations= num.iterations,
  mlsteps=fvkeeper)
}

ml.estimation<-function(y,x,cl=0,cu=1,step=1,rep=100,brk=1000){
 starter<-cs.start(y,x,cu,cl)
 s<-scorestep(y,x,starter[[1]],starter[[2]],cl,cu,step,rep,brk)
 return(s)
}

################## Alternative Regressions
#
#
#

#
# OLS
#

regression.ols.naive<-function(y,x,cl=0,cu=1){
 lsmod<-lm(y~x-1)
 return(
  list(
   coeff=lsmod$coeff,
   tvalues=summary(lsmod)$coefficients[,3],
   var=summary(lsmod)$sigma,
   r2mz=r2mz(y,x,lsmod$coeff,summary(lsmod)$sigma,cl,cu)
  )
 )
}

#
# subsample regression
#

regression.ols.selective<-function(y,x,cl=0,cu=1){
 lsmod<-lm(y[y<cu]~x[y<cu,]-1)
 return(
  list(
   coeff=lsmod$coeff,
   tvalues=summary(lsmod)$coefficients[,3],
   var=summary(lsmod)$sigma,
   r2mz=r2mz(y,x,
        summary(lsmod)$coefficients[,1],
        summary(lsmod)$sigma,cl,cu
    )
   )
```

```
  )
}

#
# Adapted Nakamura Xie estimation using the survival library
# for tobit-like estimation
#

regression.nx<-function(y,x,cl=0,cu=1){
 library("survival5")
 s<-Surv(y,y<cu)
 lsmod<-survreg(s~x[,2:dim(x)[2]],dist="gaussian")
  thissig<-var(y[y<cu])
  return(
   list(
    coeff=lsmod$coeff,
    tvalues=summary(lsmod)$table[,3],
    var=thissig,
    r2mz=r2mz(y,x,lsmod$coeff,thissig,cl,cu)
   )
  )
}
```

Appendix B

Static Model–Extensions

B.1 The validity of the first order approach

Proposition B.1 *The first order approach for model (3.9)-(3.10) is valid.*

Proof: To show the validity of the first oder approach it is necessary to show that one and only one stationary point for θ exists in the interior of $[0;1]$ and that this point generates a maximum. This leaves the possibility of a maximum achieved at the endpoints of the interval $[0;1]$. In particular, for the effort function it was assumed that $\psi'(0) = 0$. The existence of a maximum at the interval endpoints, however, is excluded if we can show that the maximizer is situated in the interior of the interval. It follows that every other value must generate a smaller value of the objective function.

The first order condition for (3.10) is

$$\frac{d}{d\theta} = sq - \psi' = 0. \tag{B.1}$$

The expression sq is constant in θ. ψ itself was assumed to increase strictly monotonically in θ. It follows that at most one θ exists which fulfills the condition (B.1). The second derivative ψ'' is positive due to the convexity assumption for ψ, accordingly $-\psi''$ is negative. Thus condition (B.1) describes a maximum. It follows that the only stationary point for the optimization problem (3.10) of the local partner is a maximum. This proofs the validity of the first order approach. ***q.e.d.***

B.2 Market structure and joint venture formation

One might suspect that the existence of a profit maximizing $s \in (0, 1)$ is a result of the market structure. One could argue, that the investor has to take the market price as given.

On the other hand, the possibility to vary the cost by choice of a joint venture arrangement gives the investor an additional degree of freedom. This perception of the joint venture makes it intuitive that under perfect competition the investor can increase his profit by changing the ownership structure a bit. However, this is the essence of this argument, a joint venture is bounded to the restrictions that limit the investor with his price choice under perfect competition. It would never occur in a market structure where the investor is free to set a price that suits him best for the technology at hand.

This idea is wrong. The decision of ownership structure is basically determined by demand characteristics. The market structure interferes with joint venture formation by the influence on equilibrium prices and quantities only. It does not influence the ownership decision directly. The following proposition demonstrates this relationship for a foreign direct investment that enters the market as a monopoly.

Proposition B.2 *Let $q(p)$ be a regular demand function. Let p^* be the monopoly price for the investor without local participation, $q^* = q(p^*)$ be the corresponding quantity and ε^* the price elasticity of demand at equilibrium.*

Let p^{**}, q^{**} and ε^{**} be the corresponding quantities for the monopoly with a local participation $s > 0$.

The investor forms a joint venture with a local partner (i.e. $s > 0$) if the following condition is met:

$$\frac{q^{**2} - q^{*2}}{2} > \frac{(1-s)p^{**}q^{**}}{\varepsilon^{**}} - \frac{p^*q^*}{\varepsilon^*}. \tag{B.2}$$

Proof: We take the derivatives for the profit functions π_s and $\pi = \pi(p)$ with respect to the prices[1]:

$$\begin{aligned}\frac{\partial}{\partial p}\pi_s &= (1-s)\left[q+(p-1)q'+2\left(s-\frac{1}{2}\right)qq'\right]\\ \frac{\partial}{\partial p}\pi &= q+(p-1)q'-qq'.\end{aligned}$$

Setting to zero and solving for the respective prices we get

$$p^* = 1+\left(1-\frac{1}{q'}\right)q \tag{B.3}$$

$$p^{**} = 1+\left(1-2s-\frac{1}{q'}\right)q \tag{B.4}$$

for wholly owned monopoly and the shared monopoly respectively. The corresponding quantities demanded are denoted $q^{**} = q(p^{**})$. Now we compute the corresponding profits and compare them:

$$\pi_0^* = \left[\frac{1}{2}-\frac{1}{q'^*}\right]q^{*2}, \tag{B.5}$$

$$\pi_s^{**} = (1-s)\left[\frac{1}{2}-s-\frac{1}{q'^{**}}\right]q^{**2} \tag{B.6}$$

where $q'^* = q'(p)\,\big|_{p=p^*}$ and $q'^{**} = q'(p)\,\big|_{p=p^{**}}$. Now we can write the profit condition $\pi_s > \pi_0$ as

$$(1-s)(\frac{1}{2}-s)q^{**2}-\frac{1}{2}q^{*2} > (1-s)\frac{q^{**2}}{q'^{**}}-\frac{q^{*2}}{q'^*}. \tag{B.7}$$

For $s \in (0,1)$, we find $(1-s)(\frac{1}{2}-s) \in (-\frac{1}{2},\frac{1}{2})$. This yields

$$\frac{1}{2}q^{**2} > (1-s)(\frac{1}{2}-s)q^{**2}, \tag{B.8}$$

thus

$$\frac{q^{**2}-q^{*2}}{2} > (1-s)\frac{q^{**2}}{q'^{**}}-\frac{q^{*2}}{q'^*}. \tag{B.9}$$

[1]Contrary to the proceedings in chapter 2 q and p do not denote equilibrium values. This is the reason for the ommission of the technology advantage A as arguments.

From the definition of the price elasticity of demand, $\varepsilon = q'\frac{p}{q}$, we get the relation $\frac{q'}{q} = \frac{\varepsilon}{p}$. Plugging this into formula (B.9) we finally achieve

$$\frac{q^{**2} - q^{*2}}{2} > (1-s)\frac{q^{**}p^{**}}{\varepsilon^{**}} - \frac{q^{*}p^{*}}{\varepsilon^{*}}.$$

q.e.d.

The importance of formula (B.2) is revealed if the figures of this condition are given the adequate economic interpretation. The quantities p^*q^* and $(1-s)p^{**}q^{**}$ are the cash flows the investor gets from his sales of a wholly owned monopoly and of monopoly with a local partner, respectively. Both quantities are scaled by their respective equilibrium elasticities ε^* and ε^{**}. On the left side we find the half of the differences of the squared quantities sold in equilibrium.

To summarize the condition in economic terms: The investor chooses a shared monopoly if the elasticity-adjusted cash flows are not too different from a wholly owned monopoly. If the difference is large the investor prefers to own the investment project as a whole.

The condition (B.2) does not require cost information to assess the preference of the investor for a shared or a wholly owned monopoly. This can be of importance for local regulatory authorities that have to deal with monopoly control.

Why should the local authorities have an interest to know if the monopoly is shared with a local firm or if it is wholly owned monopoly by the investor?

The answer is that a locally shared monopoly is expected to set lower prices than a wholly owned monopoly. This indicates that the negative welfare effect of a monopoly is less severe for a shared monopoly compared to a wholly owned one.

We find, that with a rising share the profit maximizing prices decline for a monopoly joint venture.

Proposition B.3 *Let p_i^* denote the profit maximizing price for $\pi_{s_i}(p)$. Then $s_1 > s_2$ if and only if $p_2^* < p_1^*$.*

Proof: See formula (B.4). ***q.e.d.***

This strong and, *prima facies*, surprising relationship becomes immediately intuitive if one remembers that the monopoly price is is a non-decreasing function of marginal cost under very general conditions[2]. For regular demand functions these conditions are always fulfilled. The effect of a joint venture is a reduction of marginal cost of the investor. Therefore, the monopoly price has to decrease (weakly) with a rising ownership share of the local partner.

Or, to describe it in the usual terms of marginal cost and marginal revenue, for a monopoly, the marginal cost has to be equal to marginal revenue at the optimal price. The inclusion of a local partner shifts the marginal cost curve downwards. Thus the monopoly price decreases with a rise in the local participation s.

Proposition (B.3) delivers an argument that the social cost of a monopoly is slightly lower if a monopoly is shared. Thus, the assessment if a joint venture monopoly occurs or not can be of reasonable interest for regulation agencies.

Besides the policy implications one might find that there is an important issue from a theoretical point of view. As the theory of the cournot oligopoly tells us, one can consider pure monopoly and perfect competition as the opposite extremes of an oligopolistic Cournot game with N participants where the monopoly is the outcome for $N = 1$ and the perfect competition arises for $N \to \infty$[3]. It follows, that if we find that for the extreme cases joint ventures can arise the same is true for any market structure in between.

[2]see (Tirole 1988), pp. 66-67

[3]For a proof in the case of linear demand and homogenous oligopolies, see (Mas-Collel, Whinston, and Green 1995), p. 407, Example 12.E.1. A more general treatment for the case of more general demand specifications and heterogenous Cournot oligopolies can be found in (Hoernig 2000).

B.3 Joint venture as a truth-revealing mechanism

A very attractive characterization of the joint venture arrangement can be derived, if it is considered from a contract-theoretic perspective. This perspective considers the joint venture arrangement as a contract where the local partner announces a cost reduction θ_1 for payment of[4] $\frac{s_1}{1-s_1}\pi_{s_1}$ with a cost of $\psi(\theta_1)$. It is natural to ask, if the payment according to the condition $\psi' = sq(A)$ gives the local partner the appropriate inventive. In particular, this remuneration scheme guarantees that if the local partner announces a cost reduction θ_1, he would not prefer to reduce the cost for $\theta_2 \neq \theta_1$ only.

We find the following result

Proposition B.4 *The joint venture agreement with a payment scheme according to $\psi' = sq(A)$ is truth-revealing.*

Proof: To be truth-revealing the benefits from truth-telling must be larger than those from cheating. Thus if the local partner announces a cost reductions θ_1 with an effort $\psi(\theta_1)$ but reduces the cost for θ_2 with the effort $\psi(\theta_2)$ the truth-telling[5] conditions are

$$\frac{s_1}{1-s_1}\pi_{s_1} - \psi(\theta_1) \geq \frac{s_1}{1-s_1}(\pi_{s_1} + (1-\theta_1)q(A) - (1-\theta_2)q(A)) - \psi(\theta_2)$$

and

$$\frac{s_2}{1-s_2}\pi_{s_2} - \psi(\theta_2) \geq \frac{s_2}{1-s_2}(\pi_{s_2} + (1-\theta_2)q(A) - (1-\theta_1)q(A)) - \psi(\theta_1).$$

This is equivalent to the condition

$$\begin{aligned}\psi(\theta_2) - \psi(\theta_1) &\geq s_1 q(A)(\theta_2 - \theta_1)\\ \psi(\theta_1) - \psi(\theta_2) &\geq s_2 q(A)(\theta_1 - \theta_2).\end{aligned}$$

[4] π_{s_i} denotes the profit function for the investor for a fixed local participation share s_i.

[5] See (Laffont and Tirole 1994), pp. 64-66 and proposition 1.2

But with the first order condition (3.15) this becomes

$$\psi(\theta_2) - \psi(\theta_1) \geq \psi'(\theta_1)(\theta_2 - \theta_1)$$
$$\psi(\theta_1) - \psi(\theta_2) \geq \psi'(\theta_2)(\theta_1 - \theta_2).$$

As the effort function is convex in θ the last two inequalities are true by assumption. Thus the payment scheme according to $\psi' = sq(A)$ is truth-revealing. ***q.e.d.***

Of course, the truth-telling property does provide the actual economic mechanism that yields such a result. It may be useful to sketch it.

Supposed the local partner would cheat and announce an effort level that is not planned to be provided. Then the investor could announce publicly the outcome of the partner's work and ask if some other firm could perform better. If the local partner would have be cheated this would be the case. So the investor could buy back the shares of the previous partner as long as another local firm can announce better joint work outcomes.

Since the investor always knows the results of their work he can guarantee that his profit is a little bit higher with the new partner than with the former partner. This process ends if a local firm behaves truthfully.

The truth-telling property prevents extortion once the local partner has given away a share of the investment project.

This remarkable result puts the joint venture arrangement into a new theoretical position. A joint venture is usually understood as a second-best arrangement for the participating parties due to transaction cost[6]. However, the role of a joint venture as pointed out here is that of a first-best solution for both the local partner and the investor. In other words, it is not possible to find a Pareto superior payment mode.

[6]See for example (Zschiedrich and Keller 1999)

Appendix C

Dynamic optimization

C.1 Auxiliary lemmata and results

Lemma C.1 ***Fundamental Lemma of the Calculus of Variations***
Let $G(t)$ be continuous on a closed interval $[a,b]$. Let

$$\int_a^b G(t)h(t)dt = 0. \tag{C.1}$$

for all $h(t)$, where $h(t)$ is continuous and $h(a) = h(b) = 0$. Then $G(t) = 0$ for all $t \in [a,b]$.

Proof: See (Klingbeil 1988), pp. 13-14 ***q.e.d.***

Lemma C.2 *The first order condition for the maximization problem (3.26) and (3.38) is*

$$\theta_t = s_t q_t \tag{C.2}$$

Proof: The first order condition is

$$\int_0^\infty e^{-\rho t}\left[\frac{s_t}{1-s_t}(1-s_t)q_t - \psi'\right] dt = 0. \tag{C.3}$$

To fulfill condition (C.3) it is necessary that

$$\psi' = s_t q_t \tag{C.4}$$

by virtue of the Fundamental Lemma of the Calculus of Variations, (C.1). Thus the optimization requirement for the local partner can be used without discounting. Plugging in the functional form $\psi(\theta) = \frac{1}{2}\theta^2$ delivers

$$\theta_t = s_t q_t. \tag{C.5}$$

q.e.d.

Lemma C.3 *If the integrand of a variation problem has the form*

$$F(t, \dot{A}, A) = e^{\rho t} G(\dot{A}, A),$$

then the Euler Equation yields as a necessary optimality condition

$$G_A - \frac{d}{dt} G_{\dot{A}} + \rho G_{\dot{A}} \stackrel{!}{=} 0. \tag{C.6}$$

Proof: By assumption we have $F(t, \dot{A}, A) = e^{-\rho t} G(t, \dot{A}, A)$. This yields

$$\begin{aligned} F_A - \frac{d}{dt} F_{\dot{A}} &= e^{\rho t} G_A - \rho e^{\rho t} \frac{d}{dt} G(A, \dot{A}) \\ &= e^{-\rho t} (G_A + \rho G_{\dot{A}} - \frac{d}{dt} G_{\dot{A}}) \stackrel{!}{=} 0 \end{aligned}$$

Because $e^{-\rho t} \neq 0$, $\forall t$ one can find by the Fundamental Lemma of the Calculus of Variations (C.1) that the necessary conditions from equation (C.7) are:

$$G_A - \frac{d}{dt} G_{\dot{A}} + \rho G_{\dot{A}} \stackrel{!}{=} 0. \tag{C.7}$$

q.e.d.

Appendix D

Selected Symbols and Abbreviations

D.1 Selected symbols and abbreviations for chapter 2

y	vector of observations for the response variable
y_i	response variable for individual i
X	matrix of observations of the independent variables
x_i	vector with observations for individual i
x_{ij}	observation of variable j for individual i
c_l	lower bound for response variable observations
c_u	upper bound for response variable observations
σ	standard deviation
$\phi(.)$	standard normal density evaluated at the argument
$\Phi(.)$	standard normal cumulative probability evaluated at the argument
$\mathcal{L}$	likelihood function
$\mathcal{L}_i$	likelihood function for individual i

D.2 Selected symbols and abbreviations for chapter 3

p	price
q	quantity
A	technology advantage, marginal cost advantage of the investor
MC^L	marginal cost of the local firm
$S(p)$	supply function
$q(p)$	demand function
$q(A)$	equilibrium demand
$\frac{dq}{dp}$	price derivative of equilibrium demand
p_A	derivative of the equilibrium price with respect to the technology advantage
θ	cost reduction
$\psi(\theta)$	effort function of local partner for the effort θ
s	ownership share
$\pi(p, s)$	profit function
π	profit function of the investor for a wholly owned subsidiary
π_s	profit function of the investor for a shared investment project with local participation s

Bibliography

AITKEN, B., AND A. HARRISON (1999): "Do Domestic Firms Benefit from Direct Foreign Investment? Evidence from Venezuela," *American Economic Review*, 89, 605–618.

AMEMIYA, T. (1985): *Advanced Econometrics*. Oxford: Basil Blackwell.

AMIR, R. (1999): "R&D Returns, Market Structure and R&D Joint Ventures," Discussion Paper 99/07, Center of Industrial Organization Research.

ASIEDU, E., AND H. ESFAHANI (2001): "Ownership Structure in Foreign Direct Investment Projects," *The Review of Economics and Statistics*, 84(4), 647–662.

AXELROD, R. (1984): *The Evolution of Cooperation*. Basic Books.

AZARIADIS, C., AND A. DRAZEN (1990): "Threshold Externalities in Economic Development," *Quarterly Journal of Economics*, 105, 501–526.

BALAKRISHNAN, S., AND K. MITCHEL P. (1993): "Information Asymmetry, Adverse Selection and Joint Ventures," *Journal of International Business Studies*, 24, 1–16.

BEAMISH, P., AND J. BANKS (1987): "Equity Joint Ventures and the Theory of the Multinational Enterprise," *Journal of International Business Studies*, 18, 1–16.

BERG, S., AND P. FRIEDMAN (1978): "Joint Ventures in the American Industry," *Mergers and Acquisitions*, 13(2), 28–41.

BERNDT, E., B. HALL, R. HALL, AND J. HAUSMAN (1974): "Estimation and Inference in Nonlinear Structural Models," *Annals of Economic and Social Measurement*, 3, 653–665.

BERTSCHEK, I. (1995): "Product and Process Innovation as a Response to Increasing Import and Foreign Direct Investment," *Journal of Industrial Economics*, 43(4), 341.357.

BERTSCHEK, I., AND M. LECHNER (1998): "Convenient Estimators for the Panel Probit Model," *Journal of Econometrics*, 87(2), 329–371.

BLOMSTROM, M., AND A. KOKKO (1996): "The Impact of Foreign Direct Investment on Host Countries: A Review of Empirical Evidence," mimeo, Stockholm School of Economics.

BLOMSTROM, M., AND F. SJOHOLM (1999): "Technology Transfer and Spillovers: Does Local Participation Matter?," *European Economic Review*, 43(4-6), 915–923.

BRONSTEIN, I., AND K. SEMENDJAJEV (1979): *Taschenbuch der Mathematik.* Nauka, Moskau.

CASSIMAN, B., AND R. VEUGELERS (2000): "R&D Cooperation and Spillovers: Some Empirical Evidence," Discussion Paper 2330, CEPR.

CAVES, R. (1982): *Multinational Enterprise and Economic Analysis.* Cambridge University Press.

CHIANG, A. (1992): *Elements of Dynamic Optimization*, Economics Series. McGraw Hill.

CHONG, A., AND L. ZANFORLIN (1999): "Technology and Epidemics," Discussion Paper WP/99/125, International Monetary Fund.

CHRISTENSEN, L., D. W. JORGENSON, AND L.J.LAU (1975): "Transcendental Logarithmic Utility Functions," *American Economic Review*, 65, 367–383.

COE, D., AND E. HELPMAN (1995): "International R &D Spillovers," *European Economic Review*, 39, 835–887.

CONSIDINE, T. J., AND T. MOUNT (1984): "The Use of Logit Models for Dynamic Input Demand Systems," *The Review of Economics and Statistics*, 66, 434–443.

CRAGG, J. (1971): "Some Statistical Models for Limited Dependent Variables with Application to the Demand for Durable Goods," *Econometrica*, 39(5), 829–844.

CRAMPTON, P., R. GIBBONS, AND P. KLEMPERER (1987): "Dissolving a Partnership Efficiently," *Econometrica*, 55, 615–632.

DAS, S. (1987): "Externalities and Technology Transfer through Multinational Corporations–A theoretical Analysis," *Journal of International Economics*, 22, 171–182.

D'ASPREMONT, C., AND A. JACQUEMIN (1988): "Cooperative and Noncooperative R&D in Duopoly with Spillovers," *American Economic Review*, 78(5), 1133–1137.

DASGUPTA, P., AND J. STIGLITZ (1980): "Industrial Structure and the Nature of Innovative Activity," *Economic Journal*, 90, 266–293.

DEATON, A., AND J. MUELLBAUER (1980): *Economics and Consumer Behavior*. Cambridge University Press.

DESAI, M., C. FOLEY, AND J. HINES JR. (2001): "International Joint Ventures and the Boundaries of the Firm," mimeo.

EATON, J., AND M. GERSOVITZ (1984): "A Theory of Expropriation and Deviations from Perfect Capital Mobility," *Economic Journal*, 94(1), 16–40.

EBRD (ed.) (2000): *Transition Report 2000*. European Bank for Reconstruction and Development.

FABRY, N. (2000): "The Role of Inward FDI in the Transition Countries of Europe: An Analytical Framework," Discussion Paper WP 2000-4, Universitè de Marne-la-Valèe.

FRANKO, L. (1971): *Joint Venture Survival of International Joint Ventures.* Praeger: New York.

GIRMA, S., AND K. WAKELIN (2001): "Regional Underdevelopment: Is FDI the Solution? A Semi-parametric Analysis," Discussion Paper GEP Research paper 2001/11, University of Nottingham, U.K.

GOMES-CASSERES, B. (1987): "Joint Venture Instability: Is it a Problem?," *Columbia Journal of World Business*, 3, 97–102.

GREENE, W. (1993): *Econometric Analysis.* McMillan, New York.

GROSSMAN, G., AND E. HELPMAN (1991): *Innovation and Growth in the Global Economy.* Cambridge and London: MIT Press.

GROSSMAN, S., AND O. HART (1986): "The Cost and Benefits of Ownership," *Journal of Political Economy*, 94(4), 691–719.

HARBARTH, K., T. RIEDRICH, AND W. SCHIROTZEK (1974): *Differentialrechnung mit mehreren Variablen*, vol. 4 of *MINÖM.* BSB Teubner Verlagsgesellschaft Leipzig.

HARRIGAN, K. (1988): "Strategic Alliances and Partner Asymmetries," in *Cooperative Strategies in International Business*, ed. by F. Contractor, and P. Lorange, pp. 205–226. Lexington Books, Lexington MA.

HAUSMAN, J., AND D. WISE (1977): "A Conditional Probit Model for Qualitative Choice: Discrete Decisions, Recognizing Interdepence and Heterogenous Preferences," *Econometrica*, 46, 403–426.

HELLMANN, J. KAUFMANN, D., AND G. JONES (1999): "Seize the State, seize the day: state capture, corruption and influence in transition," World Bank Policy Research Working Paper No. 2, 402.

HOERNIG, S. (2000): "Existence and Comparative Statics in Heterogenous Cournot Oligopolies," Discussion Paper ECO 2000/8, European University Institute.

IHAKA, R., AND R. GENTLEMAN (1996): "R: A Language for Data Analysis and Graphics," *Journal of Computational and Graphical Statistics*, 5(3), 299–314.

JUDGE, G., R. HILL, W. GRIFFITHS, H. LÜTKEPOHL, AND T.-C. LEE (1988): *Introduction into the Theory and Practice of Econometrics.* Wiley, New York.

KABIRAJ, T. (1999): "A Simple Theory of Joint Venture Breakdown," *Keio Economic Studies*, 36(1), 87–98.

KELLER, W. (1996): "Absorptive Capacity: On the Creation and Acquisition of Technology in Development," *Journal of Development Economics*, 49, 199–227.

——— (2001): "International Technology Diffusion," Discussion Paper Working Paper 8573, National Bureau of Economic Research.

KILLING, P. J. (1983): *Strategies for Joint Venture Success.* Praeger New York.

KLINGBEIL, E. (1988): *Variationsrechnung.* BI Wissenschaftsverlag Mannheim Wien Zürich.

KOGUT, B. (1989): "The Stability of Joint Ventures: Reciprocity and Competetive Rivalry," *Journal of Industrial Economics*, 38(2), 183–198.

KOKKO, A. (1994): "Technology, Market Characteristics and Spillovers," *Journal of Development Economics*, 43, 273–293.

KONINGS, J. (2001): "The Effect of Foreign Direct Investment on Domestic Firms," *Economics of Transition*, 9(3), 619–633.

LAFFONT, J.-J., AND J. TIROLE (1994): *A Theory of Incentives in Procurement and Regulation.* The MIT Press.

LAITILA, T. (1993): "A pseudo-R^2 Measure for Limited and Qualitative Dependent Variable Models," *Journal of Econometrics*, 56, 341–356.

LEGROS, P., AND S. MATTHEWS (1993): "Efficient and Nearly Efficient Partnerships," *Review of Economic Studies*, 68, 599–611.

LÈONARD, D., AND N. VAN LONG (1992): *Optimal Control Theory and Static Optimization in Economics.* Cambridge University Press.

LICHTENBERG, F., AND B. VAN POTTELSBERGHE DE LA POTTERIE (1996): "International R&D Spillovers: A Re-Examination," Discussion Paper Working Paper 5668, National Bureau of Economic Research.

LUCAS JR., R. (1988): "On the Mechanics of Economic Development," *Journal of Monetary Economics*, 22, 3–442.

MADDALA, G. (1983): *Limited dependent and qualitative Variables.* Cambridge University Press.

MARKUSEN, J. (1995): "The boundaries of Multinational Enterprises and the Theory of International Trade," *Journal of Economic Perspectives*, 9(2), 169–189.

MAS-COLLEL, A., M. WHINSTON, AND J. GREEN (1995): *Microeconomic Theory.* Oxford University Press.

MOOD, A., F. GRAYBILL, AND D. BOES (1974): *Introduction to the Theory of Statistics.* McGraw-Hill, 3. edn.

MÜLLER, T., AND M. SCHNITZER (2001): "Optimal Ownership Structures in International Joint Ventures," mimeo.

NAKAMURA, M., AND J. XIE (1998): "Nonverifiability, Noncontractability and Ownership Determination Models in Foreign Direct Investment with

Application to Foreign Operations in Japan," *International Journal of Industrial Organization*, 16, 571–599.

OLSSON, U. (1978): "Note on the Uniqueness of the Maximum Likelihood Estimator in the Tobit Model," *Econometrica*, 46, 1211–1215.

ROMER, P. (1986): "Endogenous Technological Change," *Journal of Political Economy*, 98, 71–102.

RONNING, G. (1990): "Share Equations in Econometrics: A Story of Repression, Frustation and Dead Ends," University of Konstanz, SFB 178, Working Paper Series II, No. 118.

——— (1991): *Mikroökonometrie*. Springer Verlag.

RONNING, G., AND K. RINGWALD (1979): "A Note on the Equality of the Constrained and Unconstrained OLS Estimates in the General Linear Probability Model," *Operations Research Verfahren*, 33, 373–375.

ROY CHOWDHURY, I., AND P. ROY CHOWDHURY (2001): "A Theory of Joint Venture Life-cycles," *International Journal of Industrial Economics*, 19, 319–343.

SAGGI, K. (2001): "Trade, Foreign Direct Investment and International Technology Transfer: A Survey," mimeo.

SCHELLING, T. (1960): *The Strategy of Conflict*. Harvard University Press.

SCHNITZER, M. (1999): "Expropriation and Control Rights: A Dynamic Model of Foreign Direct Investment," *International Journal of Industrial Organization*, 17, 1113–1137.

SIEBERT, R., L.-H. RÖLLER, AND M. TOMBAK (1997): "Why Firms form R&D Joint Ventures: Theory and Evidence," Discussion paper, Center of Economic Policy Research.

SINHA, U. B. (2001): "Imitative Innovation and International Joint Ventures: a Dynamic Analysis," *International Journal of Industrial Economics*, 19, 1527–1562.

SMARZYNSKA, B. (2002): "Technological Leadership and Foreign Investor's Choice of Entry Mode," Discussion paper, The World Bank.

STENGEL, R. F. (1993): *Optimal Control and Estimation*, Dover Books on Mathematics. Dover Publications, Inc.

STEPANOV, W. (1956): *Lehrbuch der Differentialgleichungen.* Deutscher Verlag der Wissenschaften.

SVEJNAR, J., AND S. SMITH (1984): "The Economics of Joint Ventures in Less Developed Countries," *Quarterly Journal of Economics*, 99(1), 149–167.

TIROLE, J. (1988): *The Theory of Industrial Organization.* The MIT Press.

Universität München (2002): *Database Manual for the DFG Project "Deutsche Direktinvestitionen in den ehemaligen Planwirtschaften Osteuropas".*

VARIAN, H. (1992): *Microeconomic Analysis.* Norton & Company, 3 edn.

VEBLEN, T. (1908): "Evolution of the Scientific Point of View," *University of California Chronicle*, 10(4).

VENABLES, W. N., AND B. D. RIPLEY (2000): *S Programming.* Springer.

VERNON, R. (1971): *Souvereignity at Bay: The Multinational Spread of U.S. Enterprises.* New York: Basic Books.

WHITE, H. (1980): "A Heteroscedasticity-Consistent Covariance Matrix Estimator and a Direct Test for Heteroscedasticity," *Econometrica*, 48, 483–499.

XU, B. (2000): "Multinational Enterprises, Technology Diffusion, and Host Country Productivity Growth," *Journal of Development Economics*, 62, 477–493.

YAN, A., AND M. ZENG (1999): "International Joint Venture Instability: A Critique of Previous Research, a Reconceptualization, and Directions for Future Research," *Journal of International Business Studies*, 30(2), 397–414.

ZIMMERMANN, K., AND M. VEALL (1996): "Some pseudo-R^2 Measures for Some Common Limited Dependent Variable Models," *Journal of Economic Surveys*, 3(10), 241–259.

ZSCHIEDRICH, H., AND A. KELLER (1999): *Theorie und Praxis der Markterschließung in Mittel-und Osteuropa*, no. DHS 1152 in Deutsche Hochschulschriften. Hänsel-Hohenhausen.

Thi Phuong Hoa Nguyen

Foreign Direct Investment and its Contributions to Economic Growth and Poverty Reduction in Vietnam (1986–2001)

Frankfurt am Main, Berlin, Bern, Bruxelles, New York, Oxford, Wien, 2004.
235 pp., num. fig., tab. and graphs
Schriften zur Internationalen Entwicklungs- und Umweltforschung.
Edited by Zentrum für internationale Entwicklungs- und Umweltforschung der Justus-Liebig-Universität Gießen. Vol. 7
ISBN 3-631-52208-8 / US-ISBN 0-8204-6569-0 · pb. € 42.50

During the 1990s, Vietnam experienced high economic growth, significant reduction in poverty and remarkable inflows of foreign direct investment. The book aims at examining impacts of foreign direct investment on economic growth and poverty reduction in Vietnam during 1986–2001 analytically and empirically. The triangulation methodology is used. Conclusions are that foreign direct investment contributed significantly to Vietnam's growth and this worked mainly through capital accumulation and intra-industry spillover. Economic growth in turn reduced poverty. Direct impact of foreign direct investment on poverty was insignificant at the national level but significant in three localities where most foreign investors located. Foreign direct investment thus helped reduce poverty in Vietnam.

Contents: Motivations of foreign direct investors · Channels and models for growth and poverty effects of foreign direct investment · Foreign direct investment in Vietnam: motivations and profile · Qualitative and quantitative evidences on impacts of foreign direct investment on Vietnamese growth and poverty